The Moral Weight of Ecology

The Moral Weight of Ecology

Public Goods, Cooperative Duties, and Environmental Politics

Edward Tverdek

LEXINGTON BOOKS
Lanham • Boulder • New York • London

Published by Lexington Books
An imprint of The Rowman & Littlefield Publishing Group, Inc.
4501 Forbes Boulevard, Suite 200, Lanham, Maryland 20706
www.rowman.com

Unit A, Whitacre Mews, 26-34 Stannary Street, London SE11 4AB

British Library Cataloguing in Publication Information Available

Library of Congress Control Number: 2015952407
ISBN: 978-1-4985-1453-8 (cloth : alk. paper)
eISBN: 978-1-4985-1454-5

♾™ The paper used in this publication meets the minimum requirements of American National Standard for Information Sciences—Permanence of Paper for Printed Library Materials, ANSI/NISO Z39.48-1992.

Printed in the United States of America

For His Holiness, Pope Francis I, and for my niece, Hayley—the two principal reasons for my optimism about the future of this planet.

Contents

Preface

Whenever I've had the privilege of teaching philosophy courses and assigning written, argumentative papers to students, the instructions would go something like the following. Step one: outline your argument, laying out premises and conclusions, and noting points of inference. Step two: flesh it out with any needed background information, assumptions, and matters of presumed fact that the reader would need in order to follow your thread. Step three: write it up in such a way as you understand a formal philosophical paper to read. Step four: compose a letter to your Aunt Estelle telling her what you discussed in your most recent paper and how you got your point across. Step five: submit to me a copy of the letter to Aunt Estelle. What you do with the paper drafted in step three is your business; Lord knows, I wouldn't want to slog through it.

As many a trained philosopher might attest, we often do our best work precisely when we're not attempting to "do philosophy" for our professional peers but rather when we try to make our methods accessible to everyone. This is ideally what we attempt to impart as philosophy teachers—not a mastery of pre-rehearsed arguments indexed and recalled for the proper occasions, and certainly not the ability to quote Nietzsche and sound really clever at parties (though we ultimately indulge in too much of both practices), but the ability to think on one's feet, to respond to a novel situation philosophically, and to do this by internalizing a relatively compact little set of tools that become seamlessly involved in analyzing one's own life and the world around us. Done properly, philosophy is probably the only profession that seeks to defeat itself insofar as it tries to forge a world where it is no longer necessary for anyone to busy herself consciously with the mechanics of doing philosophy. It would be better, Moritz Schlick, the founder of the Logical Positivist school, was reported to have said, that there be no more books written *about* philosophy and yet every book that *is* written be written philosophically. Ludwig Wittgenstein, the reluctant mentor of the Logical Positivists who would go on to revolutionize philosophical analysis (in the estimation of some of us, anyway), wrote that he might find it less preferable that he leave a legacy of "Wittgenstinians" than that all of us—philosopher or whatever—change our lives in such a way as to make the problems and puzzles he addressed go away.

All of this said, I still owe somebody's Aunt Estelle a very long letter about environmental politics. I would have preferred that the book you hold here be written for a more general audience that includes readers not already interested in environmental politics and not yet inclined to accept the arguments put forth either by environmentalists or by those who have serious, principled doubts about the aims of environmentalism. Aunt Estelle's likely unfamiliarity with the issues and thus her presumed neutrality might have been fertile ground for testing the quality of the cases offered on all sides of environmental disputes. She could have been our ideal arbiter, the commonsense judge of the highest court.

For better or for worse, however, Aunt Estelle doesn't exist in our world—not, at least, in the form that I conjure in my paper-writing instructions to students. "Neutrality" in our information-besotted societies is often a ruse, a reversal of what it is intended to mean. It is claimed not by the apologetically uninformed or the magnanimously impartial but often by those in denial of the ideological grip that has taken hold of them. Partisans of all stripes insulate themselves with the news and information sources least likely to chisel away at the belief system they already hold, and one they hold in part by virtue of the same self-reinforcing process of selective information-gathering. But by seizing the "center" of the already attenuated spectrum to which they've adjusted their scope, they can claim to be "neutral" on the debates and still "assessing the evidence"—all of which, of course, evidence filtered and refined to serve a decidedly partisan, nonneutral outlook.

In an attempt to counter these tendencies, this text is a bit more philosophically formal than I would have preferred it to be, and is careful to lay out arguments on all sides as charitably as possible, assessing the merits of each but ultimately settling provisionally on views that I believe survive the scrutiny. Readers more accustomed to strident polemics may be disappointed, though I've tried to present the material in such a way that those perhaps committed to one side of the existing environmental debates or another will find their positions fairly represented and thus any critiques of those positions at least somewhat compelling. Philosophy can be a painstaking business, and when we apply it to the world of politics and social relations, those pains that are taken to avoid vagueness and ambiguity, uncharitable characterizations, and hasty conclusions can indeed be difficult to bear.

The aim of the author is to make them worth bearing in the end, and I hope to have achieved this here.

Books of any sort shouldn't need to supply their own operating instruction beyond "(1) open; (2) read," but given what I've said here, and given that this text is aimed at a diverse audience of individuals with different stakes in the environmental movement and different professional perspectives on the issues (or no particular professional perspective whatsoever but merely a sincere interest), different reading strategies might be appropriate. Readers already well-versed in most of the areas addressed in this volume—environmental policy, environmental ethics and moral philosophy in general, social choice theory, basic microeconomic theory, and economic history—would do well to simply read the book in the order in which things are presented. Readers perhaps familiar with one or two of these areas but not the others, as well as readers new to this field entirely, might find it helpful to read the Introduction first, followed by the introductory and closing remarks to both Parts I and II, and only then go back and read the text from Chapter 1 forward. I've tried to confine most of the "technical" discussions to the chapters proper so that the introductory and concluding remarks can step back a bit and understand the technicalities in light of their implications for environmental policy and, ultimately, how we are to live our lives on this fragile planet. Thus previewing a snapshot describing why the more technical discussions matter might be useful in seeing the whole picture. The Afterward departs a bit from the principal lines of argument in the main body of the text, describing how the book's content (and its author) evolved over time, and can be reserved for all readers until the end, where indeed it sits.

Large parts of this work constituted, in an earlier form, a portion of my PhD dissertation at the University of Illinois at Chicago a few years back. I thus renew my gratitude to the dissertation committee and all of those who offered critical feedback on the manuscript: Sandra Bartky, Charles Mills, Ike Balbus, Neal Grossman, Richard Hudelson, and Walter Edelberg. Portions of the Introduction appeared in my article, "The Efficiency Imperative: Five Questions," in *Science & Society*, and I am indebted to *S&S* editor David Laibman and a few anonymous reviewers for their invaluable insights, as well as to the people at Guildford Press for their permission to reprint that content here. I thank also an anonymous reviewer who worked with Jana Hodges-Kluck at Lexington Press to provide me with immensely helpful comments on the more recent drafts of this manuscript, as well as Jonathan Miles at Quincy University for his thoughts on an extensively revised Introduction; readers of this volume will find it far more intelligible than it might have been had I not grappled with their questions and suggestions. Much appreciation, too, goes to Helen Aw for her editorial work on the manuscript in its prepublication forms. While I'm certain that each of these individuals would be gracious enough to share with me the responsibility for any misinformation that managed to sieve through to this end product, be assured that they warrant only credit for its merits. Factual errors, lapses in reasoning, or flights of utter nonsense that remain are due entirely to my stubbornness.

Edward Tverdek
July, 2015

Introduction

Ecology at the Margins

In 1814, the British biologist Edward Hawley reported several alarming observations to a meeting of the Royal Meteorological Society:

1. Global emissions of carbon dioxide had risen from 22.6 billion tons in 1790 to 31 billion tons in 1807—a disturbing 37 percent increase in less than twenty years. The figures translate into about 85 million tons of carbon dioxide emitted into the atmosphere each day, or about 29 pounds per person. The annual increase in emissions of about 1 percent per year in the 1790s had grown to 3.5 percent per year in the 1800–1809 period.
2. These accelerated carbon dioxide emissions had been accompanied by an unprecedented rate of tropical deforestation—about 13 million hectares per year over the same period. Given the role of forested areas in the absorption of carbon dioxide, the loss of this land was adding 6.5 billion tons of carbon dioxide into the atmosphere annually.
3. Rapid growth in previously benign areas of the globe fueled cause for concern. An 1804 report to the Academy had predicted that the United States would surpass other nations in carbon dioxide emissions by 1830. The U.S. in fact reached this mark only *two years* after the initial prediction—in 1806.
4. The noted increases in carbon dioxide emissions were leading to increased global temperatures and the associated ecological disruptions such increases entail. Conservative estimates put the anticipated temperature increase at 1.5–2 degrees Celsius by the end of the nineteenth century, with associated rises in sea level, volatility in food production, and increases in diseases associated with malnutrition. More pessimistic estimates of temperature increases of greater than 2.5 degrees by the end of the century anticipated catastrophic global changes, including the loss

of the polar ice cap. Indeed, noted Hawley, for the first time in known human history, an Arctic passage between the Pacific and Atlantic oceans had apparently opened.

The narrative above—if not the actual figures that punctuate it—is, of course, pure fiction. There was no British biologist named "Edward Hawley" of which I'm aware. There *is* in fact a Royal Meteorological Society, though it didn't convene until 1850, and its charter members had little to say about anthropogenic climate change, much less the tools to measure it accurately. I'm willing to wager, however, that you read this fictional account with somewhat more rapt attention than you might have had I told you up front that these are figures are recent estimates from a report of the Intergovernmental Panel on Climate Change (IPCC)—which indeed they are, once we add 200 years to each date referenced and change "United States" to "China."[1] Our eyes typically glaze over when we encounter such litanies in the modern setting, even the eyes of those of us inclined to crack open a book entitled *The Moral Weight of Ecology*. We've been there; we've seen that; yes, we've heard it all before, and we assent to it largely because, as Ludwig Wittgenstein described our most cherished beliefs, we hold them not as a result of sweating through the doubts but by settling into convictions for which the occasion for doubt never seems to arise or even make sense. It takes, arguably, a subversion of the usual context for such reports to retrain our focus on the details of what is being said.

I'd be further willing to wager that your interest was piqued here not merely because the context of the figures was unusual but because it was unusual precisely insofar as the narrative seeded just such doubt in your mind—a doubt about climatic claims that few of us with environmental sympathies ever experience in our current setting and which assorted nay-sayers and climate-change "deniers" wear as some sort of badge of epistemic honor. That our fictional scientist Hawley's dire predictions never did materialize by the end of the nineteenth century is a major factor in this doubt. More important, however, is the fact that they *couldn't* materialize, and we know this. Environmental harm on a scale that bothers the folks down the street is as old as our ability to produce noxious aromas and unpleasant noises; environmental harm on a global scale—the kind the IPCC figures describe—is something we instinctively associate with the economic development of in the past two centuries or so and unforeseeable prior to this modern industrial age. Granted, Edward Hawley's less fictional contemporary Thomas Malthus was peddling dismal predictions of insufficient food production, but his prognostications were based not on the damage we do but rather the tendency of human populations to outstrip an otherwise functioning natural resource base. If anything, Malthus has proven to be *under*-confident in our ability to *produce*—if not reproduce—in a way that might inadvertently have long-lasting and detrimental consequences.

This cognizance of the ecological singularity of our age is, on the one hand, quite encouraging. It betrays a collective self-awareness of our historical moment that might leave Karl Marx, a thinker with his own anticipations regarding industrial capitalism, feeling a bit smug. It has a downside, however, in fostering the belief that it's the inertia of history that is at issue in environmentalism and that we need only harness that inertia in the correct way, our values and moral convictions notwithstanding. Some free-market advocates note that innovation has two sides to it, and that while the capitalist economy may generate problems humanity hasn't had to grapple with before, it is also unique in the extent to which capitalist innovation can correct its own ill-effects in unprecedented ways and without our moral meddling; Julian Simon's (1981) *The Ultimate Resource* famously etched this notion into the libertarian collective memory more than a generation ago. Zealous environmentalists, the claim goes, fail to note this flipside of production. Leave the system unfettered, and it will recover itself. For their part, many self-described "radical" environmentalists have been no less enamored of the notion that the problems of political economy and ecology are generated within the capitalist era itself and will thus die with that era. They hold that the only sound means of addressing the environmental risks that emerge under modern capitalism require not merely the comprehensive reorganization of political and economic life but, indeed, the very way we understand "nature" and our lives within it. They have largely abandoned the ideal of a post-capitalist "economy" in the sense of the term we've inherited in capitalist societies and argue that the qualities which render capitalism unsustainable indicate that those concerns of "efficient production" that leave us environmentally strapped would be meaningless under a different, post-capitalist society.[2]

The question of whether capitalism is sustainable—whether it is doomed to excess and self-destruction or, rather, uniquely prone to resilience—is a meaningful and important question, the answers to which tell us something valuable about what we might anticipate in regard to environmental quality under capitalism. These answers are not, however, my primary aim in this book. Rather, the aim here is a normative one: to suggest how it is that certain demands upon us imposed by our environmental state may be, while burdensome, justifiable. How then does this peculiarly economic question of capitalism's fate play into these moral questions? That is the task of these introductory remarks. I will argue here that even those of us with environmental proclivities may have confused notions about just what needs to be transcended and what can be preserved in our economic relations and our relation to the natural world if we are to guarantee a sustainable future.

Specifically, I will suggest that the weighing of economic costs and benefits—and to a great extent, the concern for economic efficiency—is not the pernicious element of capitalism that put us in the environmental condition we are in, and thus that it need not be jettisoned in any sustainable economy, whatever its economic relations. At the same time, I will suggest that the evaluation of social cost of environmental harm may well tell us, as the title indicates, the *moral weight of ecology:* that is, it provides us with a meaningful way to distribute fairly the burdens of making our world sustainable.[3]

These remarks are, as we will see, thus a prelude to the principal concerns of the book. Parts I and II, following the Introduction, deal with questions that presuppose we must make some sort of value assessments of the natural world and establish our moral commitments to each other in sustaining it—as well as with the epistemic limits that shape these values and moral commitments. The basis of such value assessments are varied and contested, but the fact remains that how we value nonhuman nature must play some role in how we address environmental problems as a society and whether we even recognize them as problems *all* of us are obliged to resolve. This economic valuation of the natural world as a resource for human needs is itself neither "natural" nor inevitable. But neither is the overturning of this as a dated "paradigm" where we somehow manage to free ourselves from the requirement to assess the costs—in one set of terms or another—of what we are doing to the planet in the course of our own material reproduction. This Introduction is thus intended in part to affirm that these questions will not go away anytime soon and are indeed indispensable to environmentalism.

What a peculiar character we have in the climate-change "denier." He fancies himself a skeptic, a clear thinker versed in the ways of science, yet he is often the first and loudest to proclaim "I'm no scientist" when asked to justify his doubts regarding peer-reviewed professionals or when false humility can advance his agenda. He picks and chooses the probabilities that will govern his life with a caprice that would baffle the seasoned gambler: he deems greenhouse gas emissions harmless until all the data are in, yet takes his prescription medications promptly, even though the clinical trials that permitted their availability may have been far less conclusive than the near unanimous opinions of climate scientists on anthropogenic global warming. Even "climate" pronouncements can persuade him when conveyed as daily weather forecasts; were five media outlets to predict rain in his vicinity tomorrow while one remained "skeptical," he would not likely forgo carrying an umbrella for lack of a "scientific consensus" on the matter.

Truth be told, the genuine climate-change "denier"—and his counterpart in the "warmer," as "deniers" label those who more readily accept the claims of climate scientists—are probably more media fictions than actual persons.

We are done a great disservice in the age of the 24-hour news cycle and constant cable and internet coverage, where an overabundance of information must be packaged into familiar narrative devices and the natural environment, like most other issues, becomes a construct of those much-lamented "sound bites." For one, "climate change" has been coined to capture a host of ecological maladies—most of which perhaps traceable to our continued dependence on carbon-based fuels yet remarkably unique in how they affect local populations of human and nonhuman life around the globe. This distillation of the problem into a catch phrase has, moreover, permitted the reduction of the debates in popular discourse to the purely empirical one of whether or not it is happening: in the world of Fox News, CNN, and MSNBC, one either accepts the reality of "climate change" or one is a "denier." At its best, the discussion might be qualified to acknowledge that nearly everyone accepts the planet is getting warmer but either affirms, along with the vast majority of climate scientists, that the process is attributable to mostly human activity (and can thus be mitigated with sufficient political will) or denies, along with a handful of researchers often funded by industry coffers, the anthropogenic nature of these changes (and thus their permanence in the face of "natural" carbon cycles). This, one might further suggest, has been true since "climate change" was first touted more specifically as "global warming" in the 1980s and 1990s.

This is arguably not, however, how most informed participants in environmental debates see the possible policy landscape. Disputes over the proper response to climate change and to environmental hazards in general are not between a simple dichotomy of those who merely accept apocalyptic scenarios and those who deny them but rather among those distributed along a continuum of risk aversion. One extreme of the continuum harbors those who, while conceding that our carbon-based economy has imposed enormous problems and may cost lives, livelihoods, and economic well-being, argue that heroic measures to abate environmental harm and hasten the transition from fossil fuels could in fact be more costly than attempts to *adapt* to changes already in process. Yale economist William Nordhaus' DICE model (Dynamic Integrated model of Climate and the Economy) is perhaps the more notable among recent approaches touting this caution. Far from some affinity to climate-change "deniers," Nordhaus acknowledges potentially devastating costs if fossil-fuel emissions continue unabated but recommends a "ramping up" schedule of relatively mild curbs on emissions in the short run with incrementally more significant increases in our efforts over the next century.[4]

The opposite end of the continuum sees, conversely, these risks reversed: merely mild action on climate change and other environmental harms in the short run, those at this end argue, would in fact be far more costly than even what might strike us as drastic short-term measures. This is largely the position held by Nicholas Stern in a book-length 2007 UK report commissioned by then Chancellor of the Exchequer, Gordon Brown. *The Stern Review* offered the widely cited prediction that, while inadequate action on climate change could translate into a five percent annual decrease in overall GDP, firm action would likely result in an annual loss of only one percent and could well result in economic growth if public and private investment in renewable energy sources is done effectively.[5] Stern and his colleagues thus recommend far more strident short-term efforts by nations to reduce their dependence on fossil fuels and their resulting carbon dioxide emissions; hence the title of Stern's most recent monograph on the subject, an adaptation of recent policy lectures, *Why Are We Waiting?: The Logic, Urgency, and Promise of Tackling Climate Change* (2015).

Even—and perhaps *especially*—when these disputes are cast in the measured terms of relative risk aversion rather than the inflamed rhetoric of "denial," environmentalists of many stripes are made uneasy. Isn't it merely another form of the logic which has gotten us into this mess to render environmental policy an artifact of cold, economic calculation? Consider, for example, two specific propositions that are presupposed by the economic valuation of the natural world:

(a) The value of the natural environment is at least in part its value for human societies
(b) This value may be assessed in relations to other values—that is the *costs* of obtaining it may be weighed against other social costs

While these claims may strike some as relatively benign if not obvious, they represent for many environmentalists the abhorrent type of thinking that is at root environmental destruction itself. The first claim loses a number of points with those who would defend a "biocentric" form of environmental ethics against the "anthropocentricism" thought to pervade Western industrial societies. While I will not take up in this book the question of whether anthropocentricism is morally defensible, Part II (as well as this Introduction, to a lesser extent) will consider related questions prompted by assigning an economic value to nature; some readers may conclude from this that anthropocentrism is more or less assumed and implied, and I won't quibble with this. Suffice it to say we begin from the assumption that, even if the natural world has what philosophers describe as *intrinsic* value (I'm not disputing this one way or the other), it also has a certain instrumental value for human needs, and there is no contradiction in asserting this and calling oneself an "environmentalist."

The second claim challenges the conviction deeply held by many environmentalists that ecological concerns have proven themselves paramount and thus should be addressed without compromise and "at any cost," a sentiment expressed in the old Earth First! slogan from the movement a generation ago: "No Compromise in the Defense of Mother Earth."[6] It is this claim that I will consider here in these introductory remarks.

Among all of the implications of claim (b) considered by environmental economists, perhaps the most disturbing for radical environmentalists is the corollary claim

(b') There exists an "optimal" amount of environmental degradation.

Though some might certainly object to specifics of the wording, many economists, I would speculate, would generally agree that nothing important is lost in moving from (b) to (b'). Given, the claim suggests, that a society lacks infinite resources with which to solve its problems, there is a point beyond which it is cost-inefficient to diminish the effects of climate change, pollution, natural resource depletion, and the like. This is essentially the argument espoused by Bjørn Lomborg (2001) based on Nordhaus' DICE model. Even Nicholas Stern, with the term "urgency" punctuating his most recent title, would no doubt concede the point, however far he might situate this "optimal" point of environmental degradation from that of Lomborg. Figure I.1 portrays this point graphically. Taking the isolated and somewhat simpler example of pollution as an environmental problem, one might suggest that the "marginal social costs of pollution" (MSCP) constitute a positively-accelerating function of the amount of pollution—that is, for each additional unit of pollution, the costs to society which it amasses is a larger increment than the last unit. (There is nothing precise about the acceleration of this curve or its counterpart in Figure I.1; they are intended to clarify a point and not to conform to any particular mathematical function.) This is simply another way of saying that the worse pollution becomes, the more damaging are increases of the same amount. The reverse may be said, however, of the social costs of pollution abatement (including, perhaps most substantially, the opportunity costs of diverting such resources from other socially valuable ends: more money spent on environmental cleanup means less money available for homeless shelters). At relatively high levels of pollution, the costs of preventing an additional unit are decidedly less than the costs of preventing an additional unit when overall levels are already low, since, intuitively, it's easier to make an appreciable improvement in conditions that are foul than in conditions that are relatively tidy.[7] This is represented by the curve labeled "MSCPA" in Figure I.1.

The "optimal" level of pollution can be found at the equilibrium point of the two marginal-costs curves, point B—the point where the costs of an additional unit of pollution are equal to the costs of preventing it. Thus regulatory

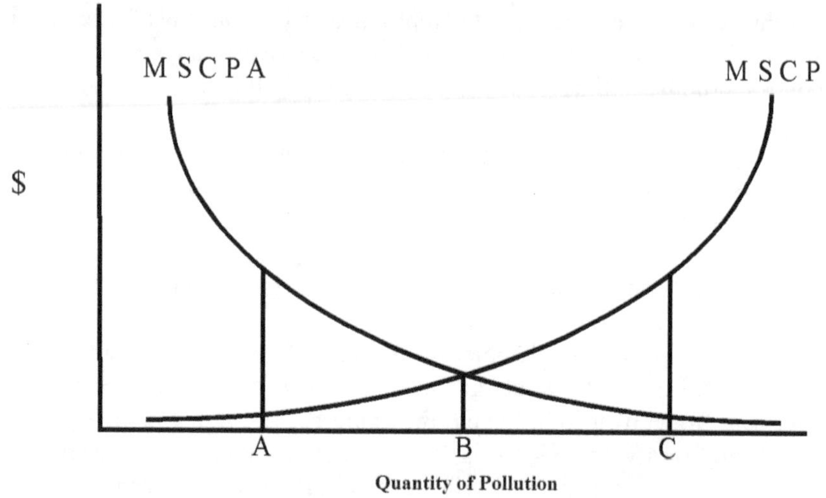

Figure I.1 Marginal Social Costs of Pollution and Pollution Abatement

pollution standards set at point C would result in "too much" pollution: the costs of an additional unit of pollution far outweigh the costs of preventing it. Just as important, for the environmental economist, a regulatory standard at point A would result in "too little" pollution: the costs of abating an additional unit of pollution vastly outweigh the costs incurred by that additional unit of pollution itself. Substitute "greenhouse gas emissions," "solid waste generation," "rainforest depletion," or whatever for the general term "pollution" (or, for that matter, any indicator of social cost for "$"), and the case remains largely unaffected.

The notion that there can be a situation of "too little pollution" or "insufficient climate change" might startle us at first, and rightly so. It is certainly not an acceptable claim for many environmentalists. Some might raise objections of the sort: "What about the harm to persons and the environment at this equilibrium point?; wouldn't the planet and its inhabitants still suffer the effects of pollution to some degree?" The answer, of course, is "yes," but it is *by definition* less harm than if greater measures were taken to ameliorate environmental degradation. Maximizing well-being is precisely what an efficiency criterion such as that advocated in (b') is aimed at; we are not really "solving" a problem if we are addressing it in such a way (or to such a degree) that the net effect to social well-being is negative.[8] We might further object that simply "managing" the costs of ecological despoliation is not nearly as noble a goal as reconfiguring the way human beings think about and interact with nature. This too may well be true; the important point is that changing the way we think about "well-being" and the scope of the interests which enter into it is a separate matter from the question of whether maximizing it is *desirable* (or, for that matter, obligatory).

Still, the environmentalist might persist, one need not endorse radical biocentrism to point out that environmental degradation is, if not "evil," certainly "misguided," and that if environmentalism is about anything, it is about eradicating such harm to the biosphere. To accept an "optimal" amount of environmental damage, the concern here, is to underwrite the same processes that have led modern societies to the brink of crisis; if we can condemn environmental damage as wrong, we must abolish it. This too, it may be argued, amounts to little more than a gallant but short-sighted ambition. To maintain it, one must be prepared to assert that the set of things which we are willing to *condemn* encompasses the entire set of things we are willing to expend unlimited resources to *prevent*, and this doesn't seem plausible. One need not submit his or her approval of something in order to acknowledge that it is difficult to abolish it entirely, and environmental economists (e.g., Heyne, 1993:29) have a viable case when they compare environmental damage to social problems such as crime in this regard. We may not like mugging, but we are entitled to judge it a crime no less when we admit that it would be prohibitively costly to prevent *every* occurrence of it, whether we measure those costs in economic terms or in terms of the violations to civil liberties that are likely to accrue in the process. So too with environmental degradation.

Perhaps one of the more vivid ways we might illustrate the plausibility of (b')—and thus the notion that it's possible to have "too little environmental damage"—is to note that the distributive effects of "too much environmentalism" can be as pronounced as the distributive effects of "too little." The environmental justice movement[9] has long held that the costs of ecological damage are borne disproportionately by the poor; those with fewer resources stand to lose more than do the wealthy if climate change is not averted or greatly mitigated. This is very likely true, but much the same would be true if we devoted more of our available efforts than is necessary toward environmental causes; we would, the economist rightly points out, be forgoing production that could have been accomplished by or on behalf of the poor to meet their immediate needs. The problem is only exacerbated over time insofar as it could foster a regressive intergenerational transfer of prosperity: if we solved all of our environmental problems while doing nothing to ameliorate the maldistribution of wealth in our present societies, we would in effect be withholding potential wealth from today's poor and reserving it for tomorrow's well-off. That this may ensure a future at all may be consoling for some, but it would meet few if any standards of *distributive justice*.[10]

Much of the reluctance of environmentalists to embrace concerns like those expressed in (b) and (b') (and, conversely, much of their allure for market advocates) stems from the understandable association of such claims with *cost-benefit analysis* (CBA) as it is practiced in U.S. and European policy-making.[11] The concern for *cost-effectiveness* is not, arguably, a call for CBA as such,[12] as even some critics of the latter have conceded, and it is not unreasonable to expect that a basic concern for assessing and minimizing the costs of laudable social policies is one to which even the most ecologically harmonious society would want to attend. Nonetheless, some environmentalists conflate the two, thus rejecting the concerns of *(b) and (b')* prematurely and with insufficient grounds. A few points concerning just where CBA departs from a general concern for cost-effectiveness thus might be instructive here.

One of the principle fears that environmentalists have of CBA is that, by assessing the relative social merits of various environmental strategies, we are "putting a price on nature," a realm taken to be "priceless" by many environmentalists. This is indeed the case with CBA, wherein *value* amounts to the aggregation of individual preferences expressed as one's "willingness to pay" (or, more generally, "willingness to assume risk") for social goods—that is, the means of valuation prescribed by contemporary welfare economics.[13] This is the currency of CBA, and it can make no claim to "work" if other measures of "value" are substituted.

Although this may appear superficially to be the implication of claim (a) above, we should note that its assertion "the environment has instrumental value for human societies" does not entail that this value be entirely subjective, as the economist may assume it to be. While it implies that nature is not wholly constituted, as I suggested earlier, by *objective* or *intrinsic* value,[14] it does not follow from this that the value it does have varies from person to person, depending upon one's psychological predispositions and "preferences." When I say that "bread has value to a famished woman," I am not making an assessment of what she is willing to bid for that bread, or any necessary reference to her psychological state. I am suggesting that bread would provide the needed nourishment regardless of any "intrinsic" value we might attribute to it, and regardless of whether the woman "priced" it or even recognized such nourishment. Claim (a) is suggesting that the natural world has, in part at least, instrumental value for human beings; it in no way requires that this stipulated value be purely subjective or expressed via one's "willingness to pay" for it (indeed, we will see in Part II that it is very difficult to maintain the belief that claims of normative value describe primitive, pre-normative consumer preferences, at least for environmental values).

While it is true, moreover, that (b) and (b') imply that any "costs" involved in environmental degradation must presuppose a measurable valuation of the environment, neither markets nor associated prices are necessary conditions for such a determination of value. What is required for cost-effectiveness is an *efficient* allocation of resources, however the value of those resources has been determined and measured. Unfettered markets are, to be sure, intended to collapse allocation into valuation; that is, the question, "How much x to produce?" is just the same question as, "How much of x are individuals willing to pay to have produced?" Yet a given society could, in principle, establish the relative value of social benefits and burdens through a variety of non-market means. Some of these may involve the aggregation of individual preferences, others may not. Some may involve democratic processes of negotiation, others may be fairly autocratic decisions of policy "experts." The point is that whatever means are used to determine the value to human beings of environmental goods and harms, and whatever we may think of those means, the question of *efficient* allocation is a different matter; the unity of valuation and allocation under absolute markets is by no means a necessary one.[15]

Even the *distribution* of environmental benefits and burdens (i.e., who bears them, regardless of the quantity we've produced) is independent of both how we go about assigning value to each. This point is demonstrated by the fact that, in particular instances, we already use "market incentives" to reduce environmental damage without the benefit of purely market means of establishing how much is "too much." The tradable emissions credits currently made available in the U.S. through the Clean Air Act revisions of 1990 (as well as worldwide via the 1992 Kyoto Protocol) attempt to do just this. The U.S. Environmental Protection Agency determines the maximum emissions allowable in select industries; rather than imposing regulatory standards on the firms within those industries, however, each is allotted a number of emission credits which it may "use" (by continuing to pollute at the level allowed by the credit) or "trade" (by reducing emissions and selling their surplus credits to firms for which it is less cost-effective to do so). While the distribution of credits at any given time is function of the emissions market, the determination of just how many credits will be made available (i.e., the maximum amount of pollution that may be produced) remains the non-market decision of a federal regulatory agency, and is based on ecological science rather than individual "willingness to pay" for clean air. The "cap" in "cap-and-trade," in other words, is a regulatory decision (enacted, economists often remind us, by "rent-seeking" legislators bolstering their own vested interests), however much "trading" ensues once that decision is made—which is precisely why many free-market zealots are not enamored of such schemes.[16] Thus we might have, as Jacobs (1994) suggests, two relatively independent applications of markets: the allocation of goods (wherein market prices reflect "value"), and the means by which a given allocation may be efficiently obtained (wherein prices may or may not constitute value). Regardless of whether one can argue cogently that markets provide the most efficient means of allocation, it is critical to point out that the determination of value—be it of environmental goods or whatever—is independent of this question, and that even the use of markets to allocate environmental goods of a given "economic value" does not require them to be "priced" via market mechanisms. CBA indeed equates market *allocation* of a good with its "value," but a general concern for cost-effective environmental policy does not require this.

Another characteristic claim of CBA that invites censure from environmentalists—one closely related to the first—is that all things of value, by being quantified, may be compared with all other things of value. We might say that CBA requires the *commensurability* of values if it is to guide policy decisions. Environmental quality must be comparable to rates of employment, which must themselves be comparable to public health and safety, which must be comparable to productivity, and so on. Advocates of CBA can assert this because, as we have just seen, CBA assess the value of things in the world in accordance with the welfare economist's measure of one's "willingness to pay" to satisfy one's preferences; however intangible a good might be (and environmental goods can be fairly intangible), it can be priced for CBA by aggregating individual preferences expressed and quantified in these terms.[17]

As one might anticipate, the claim of commensurability is subject to a good deal of criticism, not the least of which from environmental advocates (e.g., O'Neill, 1993). It is sufficient for our immediate purposes to again merely note that while commensurability is necessary for CBA, it is not required to advance claims like (b) and (b'). This may not be obvious, given the wording of (b), which asserts that environmental values may indeed be compared to other values (and thus the respective costs of obtaining them compared) as well as (b'), which asserts that there is thus an "optimal" amount of environmental integrity we might aspire to. What we have already said of claim (a), however, extends to claim (b): it is indifferent to the question of whether the value of the natural world can be determined by aggregating individual preferences (much of the balance of this book is intended to argue that it cannot be so determined). Since it is not wedded to a singular measure or source of value, the claim that some things of value may be "weighed against each other" thus seems fairly non-controversial. Translated into prescriptions for social policy, it merely suggests that *among* those things which we deem—by virtue of sound moral argument, free democratic process, etc.—appropriate to compare, we would do well to allocate them efficiently; for example, if environmental quality is justifiably comparable to the socioeconomic well-being of human populations, we would want the highest level of such well-being as is compatible with the highest level of environmental quality.[18] This does not assert that environmental quality can be weighed against *all* other things of value, or even that it is unequivocally comparable with human well-being. It asserts only that if a plausible argument could be made with respect to the latter, it would be irrational to disregard it for the sake of asserting the singular importance of environmental quality, or any particular value.[19] The fact that commensurability of all values may well be indefensible does not suggest that *no* values are comparable nor compel us to abandon all attempts to make such comparisons.

Let's summarize the story so far. Environmentalists critique what they perceive to be the "calculus" of capitalism, and suggest that the desire to weigh costs and benefits of environmental policy merely reiterates the problem of our jaundiced understanding of the natural environment as a resource for human exploitation. I have suggested in these most recent pages that, whatever the merits of a more "biocentric" understanding of human and nonhuman nature, environmental policy ought to remain "rational" in the sense that we recognize the trade-offs we face and that we choose wisely. But the recalcitrant environmentalist still has a case to make regarding the economic efficiency that is thought to be part and parcel of capitalist development. He or she might suggest that it is not the accounting itself that is the problem, it is *the stuff that is counted*—specifically, the sheer abundance of it and our difficulty in disposing of waste and excess generated in the process of production. Capitalist development, this line of reasoning suggests, is the source of our environmental problems not merely because the calculus of costs and benefits compels us to treat nature as a warehouse of human satisfaction but because it *fills* that warehouse with commodities and the unwanted and unusable by-products of manufacture. It is to this concern these introductory remarks now turn, and to do so, we need to retreat a bit to the basics. If the claim under consideration is that precisely what capitalism does best is the potential source of its demise, we should be clear on what it is thought to do best.

I have acquired, through my own labors, seven pounds of grain.[20] I need only five. Similarly, you have acquired ten fish and need only six. We agree to barter our respective surplus of grain and fish and are both made better off in the process; I now have sufficient grain *and* some fish, while you have added grain to your piscine diet that you would otherwise not have. Moreover, we both have more grain and fish than we would have had we tried to produce both ourselves.

On a typical day, households in the U.S. alone will spend $84, with $26 of that devoted to housing, and $12 going toward food. On that same day, the New York Stock Exchange will see over 250 million shares of stock change hands, with a total dollar value in the billions.[21] Compounded with these transactions are numerous buyouts and mergers which, while perhaps not a daily spectacle, concentrate billions of dollars in assets in the hands of an increasingly narrow range of telecommunications, pharmaceutical, energy, and similar firms. In the banking industry alone since 1994, ten of the largest banks were reduced to five via mergers, with a total of nearly $19 billion paid out by the acquiring firms, increasing their assets by an average of 53 percent or $28.5 billion.[22]

The first situation, while typical of classical economic literature, is likely to strike one as pedantic; we can grasp it fairly easily, though we may wonder what economists gain in inviting us to do so (note that such exchanges sound more simple when they involve "primitive" goods like grain and fish; we may just as well have bartered bagels and portions of lemon-broiled halibut at a loss of little more than rustic charm). The second situation, while more relevant to the daily workings of capitalism, seems unfathomable to many of us. The account makes sense, insofar as one knows the meaning of words like "stocks," "mergers," and "assets," but a picture of the net business being discussed is, like Descartes' chiliagon, not something that one can seize in one's mind.

For market advocates (and even some critics), however, the latter situation makes sense precisely because of the former, insofar as the two states are points on a continuum. But how to characterize this continuum? What is it exactly that links complex capitalist economies and simple two-person exchanges? One prominent candidate is *reducibility*: however complex advanced capitalism may become, it remains, at its normative root, the sum of a finite number of two-person, mutually advantageous transactions.[23] Stripped of the economists' vocabulary, this is a fairly familiar notion. Indeed, it comprises the crux of contemporary defenses of market capitalism which ask us to "work backwards" from a dense, hypercomplex economy toward the elementary exchanges which comprise it. The agreeable absence of anything which could be construed as coercion in the simple case, we are told, gives us grounds for concluding that there is nothing morally reprehensible about the capitalist economies that evolve from it—including and especially the distribution of wealth that results, however disparate its poles may be. This is the stuff of libertarian diatribes against egalitarian "patterns," and was codified elegantly by Robert Nozick (1974).

In addition to this backward-looking claim of normative reducibility, one might point to the same presumed continuum between simple and complex exchanges to make a non-normative, forward-looking claim: market transactions preserve, in the intricacies of capitalist economies, the *efficiency* which obtains in the simplest, two-person exchanges.[24] We are familiar with this assertion largely through its most famous articulation: Adam Smith's account of the "invisible hand" at work in capitalist enterprise. It is a single, brief passage culled from nearly five hundred pages of text and cited with a ubiquity that would no doubt amuse Smith today, given its context (his concern at this point in *The Wealth of Nations* is with the relatively mundane issue of tariffs and prohibitions placed upon international trade). Still, it remains a passage well worth reproducing here:

> As every individual, therefore, endeavours as much as he can both to employ his capital in the support of domestic industry, and so to direct that industry that its produce may be of the greatest value; every individual necessarily labours to render the annual revenue of the society as great as he can. He generally, indeed, neither intends to promote the public interest, nor knows how much he is promoting it. By preferring the support of domestic to that of foreign industry, he intends only his own security; and by directing that industry in such a manner as its produce may be of the greatest value, he intends only his own gain, and he is in this, as in many other cases, led by an invisible hand to promote an end which was no part of his intention. Nor is it always the worse for the society that it was no part of it. By pursuing his own interest he frequently promotes that of the society more effectually than when he really intends to promote it. I have never known much good done by those who affected to trade for the public good. It is an affectation indeed, not very common among merchants, and very few words need be employed in dissuading them from it. (Smith, 1976 (1776):477–478)

While Smith himself had little reason to suspect that the phrase "invisible hand" would be permanently tethered to his name via generations of high-school history quizzes, his point would become a watershed in the transition from Classical to Early Modern political philosophy. We cannot, Smith counsels, survey all of the transactions of a complex economy with any degree of competence—at least not in such a way that would reveal the true interests of society at large. We can, however, observe the outcomes of our own transactions with others and extrapolate from them. If you and I can be described as "better off" by virtue of voluntary exchange, there is no particular reason to limit that quality to *our* transaction. It is a mere node in a sprawling network of similar dealings, each of which, owing to its observable ability to maximize private interests, translates into the interests of society writ large once summed. Smith's metaphor thus remains, arguably, the basis of the modern economist's belief that *efficiency* is a purely descriptive rather than normative concept and that altruism, however commendable, may be essentially superfluous in ensuring the general economic well-being of society. With the emergence of the invisible hand, we abandon the Aristotelian and Scholastic conceit that "the good life" could be divined by mortals and make it a tidy matter of definitional fiat. The "good" is simply what emerges from everyone pursuing his or her own ends, and we know the inscrutable aggregate to be "good" from the quality of its elements.

It is this last claim that leaves many contemporary critics of capitalism especially puzzled. Why, they ask, should we consider *efficiency* a purely descriptive concept if we are talking precisely about our ability to maximize what we produce and thereby judge people to be "better off" than they would be otherwise? Isn't this aim a moral matter?

The typical response offered here by market economists is that *efficiency* refers not to the quality or justifiability of our selected ends but rather to the best and most effective strategy for achieving them, whatever they may be. *Efficiency*, in other words, answers "how?" questions, not "what?" or "why?" questions; it gives us at most a prudential description about how to achieve ends we've already chosen ("if you want *x*, do *y*"), not a prescription as to what those ends ought to be.

This answer, though, rests largely on a semantic ploy. It exploits an ambiguity between "efficiency" as a selection criterion for choosing among means to given ends and "economic efficiency" as a policy end in itself. When the economist advises us to maximize the social product with the fewest possible inputs—something that arguably most economists are advising indirectly if not directly—she most decidedly *is* advocating particular ends. This is precisely what results from being efficient; we have more of something at the lowest possible expense of effort, resources, etc. For her to protest that she is really only talking about means rather than ends is either insincere or naïve: she has pre-selected exactly those ends that comport to the means of efficiency. If our ends are simply *having enough x for everyone* or *having more x than p*, we might settle for something other than efficient means of getting there. If our ends are indeed *maximizing x* per unit of input, however, there is little doubt what sorts of methods we will adopt to achieve them: the most *efficient* ones. To adopt any other means would, by definition, yield less of *x* than we might have yielded otherwise. Yes, deciding *which* methods are indeed the most efficient in achieving that goal is undeniably a descriptive question. But the economist is pulling a quick switch here. She is suggesting that, because one's preferred method for a policy of maximizing the social product per unit of input is subject to purely empirical tests, there is nothing prescriptive about the policy itself. This neglects the fact that "maximizing the social product per unit of input" has been embraced over other aims—a choice that can only be defended on normative grounds.

The misdirection involved here is probably most visible yet often unchallenged when economists retreat to the trivially true claim that *efficiency* is not a matter of *distributive justice*. The latter, we are told, is concerned with a *fair* distribution of the social product—a product that may or may not be maximized by efficient means; that efficient outcomes maximize output per unit of input says nothing, the argument goes, about how to divvy it up. *Efficiency* is simply about making more products available for the dividing, not the division itself. Even in the more restricted case of *Pareto superiority*, no one, by definition, is losing out in achieving that state, and at least one person is improving her lot. Sure, not everyone gains, but pursuing economically efficient outcomes ensures that no one is really worse off than she was before. No harm—as the saying goes—no foul. And where "fouls" can't thrive, normative concerns are inapplicable.

The fact that any distributive ends can be pursued more or less *efficiently*, however, should not obscure the additional fact that "maximizing the total product at the least cost without regard to its distribution" is no less prescriptive in its content than any other proposition regarding how we ought to produce. We might, for example, be so beholden to the notion of *equality* that we are willing to level the social product for its sake, benefitting the dispossessed, no doubt, but also disadvantaging those who currently possess a great deal. We might, in fact, be so smitten by *equality* that we're willing to do this in a hypothetical society where even the "dispossessed" are already relatively comfortable and those whose pockets from which we liberate the wealth see a substantial decline in their well-being. Would doing so be *wrong*? It might well be. The point is that it can only be pronounced "wrong" on normative grounds. Translating "wrong" here as simply another descriptive term for "inefficient" won't (and shouldn't) stop the conversation; the recalcitrant egalitarian's aim is not maximization of the social product via the route of fewest inputs, let alone the benefit of at least one person amid costs to no one. His aim is equity. At best, he may be concerned with the most "efficient" way to distribute the product—whatever its magnitude—as equitably as possible, such that, *for example*, adding a unit to anyone's lot is detracting first from those with possessions in greatest excess of that lot. But this is just another way of saying he wishes to maximize equality per unit of effort rather than, say, ensuring *sufficient* equality or maximizing something else. To forgo the goal of equity via *some* means for a goal of maximizing the social product through the most *efficient* means is not to forgo normative ends; it is merely to adopt different normative ends.[25]

If mainstream economists—arguably capitalism's most vocal advocates—are prone to overlook the normative content in the efficiency imperative in capitalist growth, they are not alone. Some of capitalism's critics take the maximization of output to be a descriptive function of the system's inner logic rather than a choice of ends that comes with a particular economic program attached. Proponents of radical environmentalism roundly criticize capitalism for its perceived tendency toward economic expansion, and its consequent inability to accommodate the "limits to growth." Typically, this propensity is attributed to the characteristic of capitalist production in question above—one which distinguishes it from the economic systems which preceded it and, presumably, from those systems with which a "deep" environmentalist would supplant it: the production of commodities for exchange rather than goods for use.[26] This is not a normative criticism about the justifiability of particular economic exchanges—however much radical environmentalists may hope to draw from it a moral condemnation of capitalism. Nor is it a claim that capitalist production is inefficient in maximizing the net social product; indeed, it would seem that the radical environmentalist is suggesting capitalist production is all "too efficient" in this regard. It is, rather, a further descriptive claim about the course of capitalist development, and yet a third way in which we might understand the continuity between simple exchanges and full-blown capitalism: the kernel that motivates specialization and exchange (and which may well be innocuous) in the simple case leads historically to the current case of superfluous commodities and profligate consumers. This notion has arguably characterized a substantial portion of the environmental movement for the last generation or so. Thirty years ago, British environmental activist Jonathan Porritt, for example, described this as an "insidious contradiction in the way the [capitalist] system seeks to meet people's needs":

> At the heart of the problem is the fact that wealth is basically distributed through the jobs people do. The only reason why many people do the jobs they do is to earn an income. As well as producing the things we do need, this means that many things are produced which we don't need. People produce junk and cajole others into buying it, purely as a means of earning an income. If the system works (i.e. we achieve full employment), we basically destroy the planet; if it doesn't (i.e. we end up with mass unemployment), we destroy the lives of millions of people. (Porritt, 1985:45–46)

A few years later, Carolyn Merchant offered a similar pronouncement on the nature of capitalist growth:

> A major problem confronting the capitalist system is the inherent necessity for economic growth. Capitalists make money for further expansion by creating products that consumers will purchase. They do so by fabricating needs for more and fancier food, clothing, and homes, as well as producing luxury items such as better cars, television sets, video recorders, electric shavers, blenders, and microwave ovens. Why not stop the growth mania and focus on quality of life items that fulfill basic needs? If any given producer curtails growth, she or he will be bought out or forced out of business by a competitor. If all capitalists agree together to curtail growth, massive unemployment will occur in a system in which population continues to grow. (Merchant, 1992:25)

Capitalist economic development, radical environmentalists have long held, carries with it structural limitations to its own sustainability: it must either continue to produce the superfluous commodities that will ensure its environmental breakdown, or it must curtail such production, thereby ensuring its economic stagnation.[27]

Why is it that this was never an issue prior to capitalism? Even though our historical hindsight suggests that precapitalist societies didn't have the productive capacity to face the dilemmas that Porritt and Merchant attribute to capitalism (the very hindsight that led us to find a little odd that fictional 1814 report of the Royal Meteorological Society), could not it ever have occurred to these societies to situate themselves as particular economic forms and assess their own sustainability? If so, why are we any different? Why, in other words, should the claim that there is something endemic to current productive arrangements that makes them incapable of sustaining themselves carry more weight at the beginning of the twenty-first century rather than, say, the late fifteenth century, or just prior to the Neolithic revolution ten thousand years ago? This would seem to be a relevant question if one were doubting the environmental sustainability of capitalism, the answer to which may not be as simple as one might initially believe. There have been, undoubtedly, periods of relative scarcity of resources throughout history, as well as crises in the treatment of "waste" in the production process (including that of humans—recall the plagues of medieval Europe). While species extinction has certainly accelerated in the past few centuries, tens of thousands of species had been extinguished with surprising regularity throughout history—many of which through predation. Why were these environmental trends not just as plausibly called upon to demonstrate the "unsustainability of feudalism," or the "blight of human agriculture"?

What we are asserting when we attribute this self-examination as something unique to capitalism, arguably, is that capitalism has a trajectory—a tendency toward its own demise that marks it as peculiar among economic arrangements. Our expectations for this tendency may be bolstered by the accumulation of environmental data, but it is knowledge we ought to have even in the absence of confirmatory evidence. It is a tendency, moreover, that reveals its own particularity such that inhabitants of capitalism are uniquely concerned about its sustainability, since it is only capitalism (in comparison to its historical predecessors) that is, this assertion contends, unsustainable by design. Thus while even capitalism's advocates will acknowledge that complex capitalist economies can be reduced to their simplest exchanges—either by virtue of their normative justifiability or their efficiency—it is only this third account which suggests that the essential seed of capitalism is also the "seed of its own destruction."

If this project sounds familiar, it shouldn't be surprising. Identifying the characteristic and potentially self-destructive flaws of capitalism has been largely the domain of Marxism for the last century and a half. Despite their general antipathy toward Marxism, radical environmentalists advance a number of arguments as to the unsustainability of capitalism that rely on the tools and language of historical materialism. Porritt and Merchant, for example, both allude to a fundamentally Marxist/Hegelian notion of *contradiction* to describe the accumulation of incompatible tendencies in capitalist economies (indeed, Porritt makes explicit use of the term). They argue that there is something unique about the modern period of capitalism in that it fosters economic growth in the production of commodities, and any such system must either continue to produce expansively or face destabilizing economic stagnation.

Not all environmentalists, however, harbor this dismal view of capitalism. Indeed, some are quite optimistic about the economic prospects of status quo once the developed world weans itself from a fossil-fuel-based economy and embraces renewable energy sources. Nicholas Stern, for example, whom I earlier noted as an exemplary voice on behalf of more stringent emissions limits in the short run and a generally more robust environmental policy, remains notably confident that such measures would in fact be economically beneficial: "There are indeed," he suggests, "paths that combine economic growth with climate and environmental responsibility."[28] This optimism is not entirely new to the environmental movement, and can arguably be traced back to the "green entrepreneurialism" professed by authors such as Paul Hawken (1993) since the early 1990s. It is more than the trivially true observation that, if stringent carbon emissions limits are put into place in a given region, the market for alternative energy sources would be primed; the claim is that the world economy can flourish on non-carbon energy sources while still mitigating the other social problems we address in the current, largely carbon-based economy, such as poverty, access to health care, and the like. While advocates of this position can hardly be called "free-market capitalists"—and, indeed, many of their libertarian and antienvironmental critics point out they are calling for extensive government intervention into the economy in the form of steep carbon taxes to fund generous subsidies for alternative sources—they can be said to retain a good deal of confidence in traditional notions of competition and innovation, the perceived drivers of capitalist technology.

Between these two extremes—that is, between (a) those radical environmentalists resigned to the notion that capitalism has lived its useful life and that any environmentally sustainable society must be post-capitalist and potentially zero- or negative-growth, and (b) those who envision continued growth via technological solutions and a thriving market economy nudged out of its dependence on fossil fuels—sits what I would speculate is the larger contingent: those of us sympathetic to the cause of environmentalism who have taken relatively modest steps to lessen our personal impact on the planet yet who haven't really considered the economic implications of moving beyond the fossil-fuel industrial age, much less the potential threats to our own relatively comfortable lifestyles. We recycle, of course, and we've cut down on our use of disposable packaging. We get much of our produce at the local farmer's market. Perhaps we've even purchased a hybrid vehicle as well as energy-efficient appliances and a solar-heated shower unit. That's about it, right? With enough political pressure, someone will come along and change my principle energy sources from fossil fuels to renewables, but I will basically stay on the grid doing pretty much what I'm doing now, right? I can still purchase clementine oranges in Chicago in January and take overseas trips now and then, right?

Not necessarily. As the journalist and environmental activist Erik Lindberg is fond of pointing out, Western "liberals" who fancy themselves "environmentalists" often succumb to the myth that their lifestyles will remain essentially intact provided that nations take immediate action to transition from fossil fuels to renewables. This cannot be, Lindberg and others point out: our current levels of consumption in much of the developed world have been conditioned by an energy history on a geological scale that cannot be replicated in technological time: tons of organic material processed by time and by gravity (it takes half a billion years under intense pressure to turn organic material to oil) and at no particular cost to humanity long before we arrived (Lindberg, 2014). Alternative energy sources and green technologies may be indispensable to a sustainable future, these environmentalists concede, but they will not serve as a panacea, and they will be incapable of sustaining our current levels of consumption.

That even in the most optimistic scenarios the prototypical American consumer lifestyle may be significantly altered is not news to radical environmentalists of the past generation. Voices long associated with the environmental movement such as those of Herman Daley (1991) and Juliet Schor (2010) have praised the environmental and personal benefits of a simpler lifestyle and a "steady-state" economy based on quality of life rather than endless growth, all the while expressing skepticism of technological, clean-energy "solutions" that purport to maintain our extravagant consumer lifestyles with benign effects.[29] By the beginning of this century, the Intergovernmental Panel on Climate Change (IPCC, 2001) was calling for, somewhat more prosaically, individuals to enhance their "personal resource productivity"—that is, their "ability to maintain and/or increase satisfaction with lower and/or intermediate input of resources"—and suggesting that an "intervention in the prevailing narrative of consumption . . . [may be] a possible strategy to interrupt the satisfaction–consumption cycle."[30] Even Pope Francis, in his long-awaited encyclical on the environment, aligned himself with this position; *Laudato Si* contains no less than 64 references to cognates of *consume*, invoking the philosophy of Romano Guardini to condemn both the "obsession with a consumerist lifestyle" (#204) and the "blind confidence in technical solutions" (#14).[31]

Why is it, then, that many of us in the developed Western world prone to support policies that curb emissions and bring the fossil-fuel economy to an end as soon as is feasible tend to anticipate our lives as continuing largely as they are, with reliable comforts and predictable commodity production? How do we reconcile a call for strident public policies and profound political change amid the expectation that our consumption habits will emerge largely unscathed? One possible reason may be the relative ease of abstraction in an era when the public face of the environmental movement has migrated from largely a local- and lifestyle-based political force (recall that the worldwide Green movements of the mid-1980s often adopted the second-wave feminist slogan, "the personal is political," in addition to forging their own: "think globally; act locally") to one where much of the focus is on international agreements (Kyoto, Rio de Janeiro, etc.) the domestic policy effects of which—including the consequences to personal consumption habits—might be years away.[32] Many, to be sure, have made drastic changes to their lives intended to cultivate a smaller carbon footprint. But many more who might find this laudable and who might anticipate such changes in the future have yet to consider that, for example, the rates of consumption of meat and other livestock products that we have now could not likely persist in a world on an ecologically sustainable path.[33] Are most people who are genuinely concerned about climate change also genuinely prepared to abandon steak dinners, egg-and-sausage breakfasts, and milk on demand?[34]

Given the complexities here, it is perhaps not surprising that there exists ambivalence among even those in the economically developed world who claim affinity to the "environmentalism" mantle. Once we put aside the contrived and grossly oversimplified "warmer/denier" narrative, we find an environmental advocacy landscape of considerable nuance. Some, such as Nordhaus and Lomborg, see the problems of climate change and its associated maladies with a sincere eye toward mitigating them, but point out that international agreements (and the resulting domestic policies) aimed at cutting greenhouse gas emissions and hastening the demise of a fossil-fuel-based economy can have substantial costs—not just to the coffers of firms with a vested interest in carbon-based fuels but to everyone who receives remote benefits from the industrial economies of the world, including the impoverished who have grown dependent on corporate largess for their meager existence. Others—most notably Stern in recent years—have a more urgent reading of the situation, and are thus more willing to accept the temporary economic costs imposed by stringent emissions controls and steep subsidies for alternative energy sources with the view that the long-term balance sheet will fare better than it would with more cautious approaches. Overlaid on this dimension of urgency and risk aversion, however, is quite a different issue: how confident are we that a growing economy of consumer goods can persist in a post-fossil-fuel world? It is notable that this question is largely orthogonal to the first, and that even those policy advocates who disagree on the urgency question might well be aligned on their optimism for the ability of alternative energy sources (with a great deal of extra-market coordination and support) to maintain a growth economy, while even some who advocate, along with Stern, swift and decisive action on climate change have little hope that alternative technologies will somehow maintain a growing consumer economy, and thus counsel that we learn to appreciate a simpler lifestyle more appropriate to a post-fossil-fuel world and less dependent on consumer goods. Figure I.2 maps this two-dimensional advocacy landscape graphically:

In the upper left-hand quadrant sit, by my estimation, those who are low on the "urgency" dimension but high in their confidence that, transitioned to properly and perhaps with some regulatory coordination, a post-fossil-fuel consumer economy can be as expansive as the current one. Call them "skeptics," in reference to Lomborg's (2001) self-identification (and recall that they do not *deny* the threat of anthropogenic climate change but merely doubt the wisdom of immediate, heroic policy measures). In the upper right-hand quadrant, on the other hand, are those who advocate a more aggressive environmental program, perhaps at the short-term risk to other social ends deemed valuable. These "techno-greens" or "eco-entrepreneurs," if we may call them that, are just as confident as their counterparts at the skeptical end of

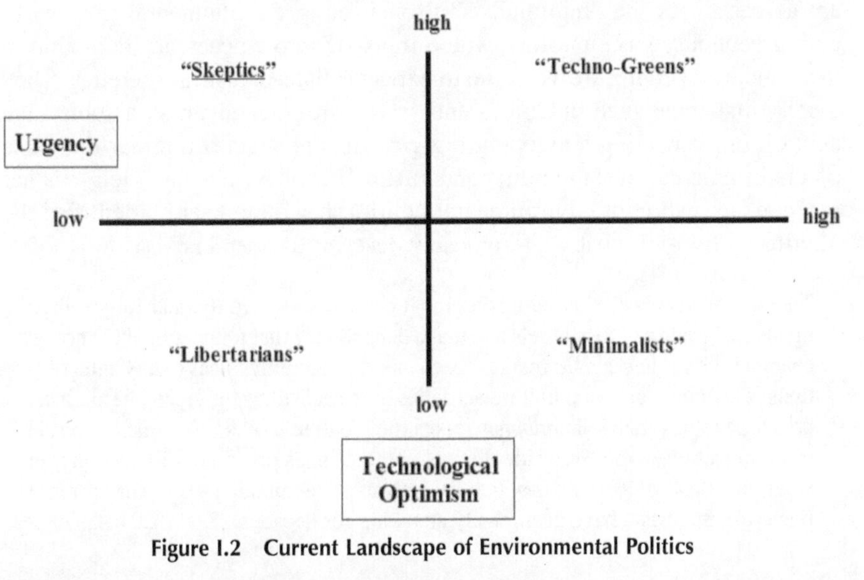

Figure I.2 Current Landscape of Environmental Politics

the risk dimension that an ecologically sustainable economy is one that may well include steady growth and the continued mass production of consumer goods. Nicholas Stern, I believe, could rather incontestably be situated here as a current exemplar of this view, though there are no doubt others that could be positioned here. Below them, in the lower right-hand quadrant, are situated those who, while similar to the "techno-greens" in urging immediate and decisive action to arrest greenhouse gas emissions and avert serious climate problems, welcome the attendant demise of capitalism's consumer culture, wasteful habits, and environmentally demanding lifestyles. Call these, for shorthand purposes here, "minimalists" in reference to the movement that has adopted this label. Exercise caution with this loosely unifying label, however: not all policy denizens of this quadrant who are committed to simpler lifestyles are necessarily motivated by environmental concerns, and they are joined here by advocates of positions similar to theirs but as diverse as those of "Green" Marxists and, arguably, Pope Francis.

What can we say of those in the lower left-hand quadrant? What do they espouse? What does it mean in the current environmental advocacy landscape to be simultaneously averse to regulations that would lighten our environmental impact on the globe *and* convinced that such measures would effectively stymie rather than foster the capitalism we have now? Inhabitants of this political space, I think, are about as close as we come to the fabled climate change "deniers" of the contemporary news-cycle narrative. Their "denial," however, amounts not literally to the principled rejection of a scientific hypothesis based on counterevidence but rather doubt as to the *authenticity* of the environmental movement at large. Those in this contingent are as certain as the "minimalists" that enacting environmental policies to curb greenhouse gas emissions would translate into a necessary reduction in the standard of living we've come to expect of late-industrial societies. They merely disparage such outcomes and see environmentalism as a subterfuge bent on implementing a creeping big-government socialism under the guise of environmental sustainability and rustic lifestyles. Naomi Klein, in her recent examination of environmental politics, has quite keenly identified the ulterior motives of what we too loosely describe as "deniers":

> They know very well that ours is a global economy created by, and fully reliant upon, the burning of fossil fuels and that a dependency that foundational cannot be changed with a few gentle market mechanisms. It requires heavy-duty interventions: sweeping bans on polluting activities, deep subsidies for green alternatives, pricey penalties for violations, new taxes, new public works programs, reversals of privatizations—the list of ideological outrages goes on and on. Everything, in short, that these think tanks—which have always been public proxies for far more powerful interests—have been busily attacking for decades. (Klein, 2014:39)

These ideologues may be, Klein continues, sorely misguided on their understanding of climate science.

> But when it comes to the political and economic consequences of those scientific findings, specifically the kind of deep changes required not just to our energy consumption but to the underlying logic of our liberalized and profit-seeking economy, they have their eyes wide open. The deniers get plenty of the details wrong…but when it comes to the scope and depth of change required to avert catastrophe, they are right on the money. (Klein, 2014:44–45)

Call this quadrant "libertarian" (in the strictly economic sense, rather than, say, those who advocate civil liberties in general), insofar as their unifying concern is the preservation of a largely mythical capitalism free from government intervention and the imposition of particular political values on society at large by those unwilling to submit their preferences to the arbiter of the free market. This isn't to say that "libertarians" as I'm labeling this position here are necessarily opposed to environmental quality.[35] Their point, rather, is that the allocation of any good—be it environmental quality, tablet computers, or athletic tube socks—should be calculated not by scientific research or one's moral values but rather, as we've seen, by the meeting of what manufacturers are willing to produce and what consumers are willing to pay for that product; they are thus as suspicious of what I've labeled as "techno-greens" as they are of "minimalists," since the thriving economy that environmental technophiles envision is one brought about by substantial government intervention and taxpayer subsidy of alternative energy sources. This is the stuff of the Cato Institute, the American Enterprise Institute, and similar self-identified libertarian think tanks. Whether we see them as a small fringe group of right-wing zealots or a pervasive stealth influence on policy-making in the Western capitalist world, I would wager that, as environmental treaties and public policy start to make increasing demands on consumer habits, more and more individuals will come to find themselves sympathetic to their cause, even those who profess environmental sympathies. It is thus toward the arguments of this group that the balance of this book is aimed.

Even if only some of the predictions of climate scientists prove to have been accurate, and even if the nations of the world implement with haste those policies needed to soften the adverse effects of climate change that do indeed manifest themselves, this would likely have substantial implications for the way we conduct our lives. Further agreements would need to be sealed at the international and national levels, cooperative plans would need to be organized, duties to adhere to them would need to be enforced, and many of the planet's human inhabitants would be compelled to accept a standard of living perhaps lower than the one they currently enjoy (or endure, as the case may be). Each of us—even those individuals indifferent or even hostile to environmentalism—would need to act as if we placed a great deal of value on the natural world. Should we find these prospects disturbing?

My aim in the remainder of this book is not to navigate the policy debates outlined in this Introduction, be they between advocates of urgent action on the environment and those counseling a cautionary, "ramping up" approach, or be they between environmentalists who anticipate an economy thriving as a result of its transition to renewable energy sources and environmentalists who concede consumer capitalism and the lifestyles it breeds are unsustainable and that only by abandoning them can we hope to undo what we have done. Rather, my aim will be to consider the moral implications of even modest ecological change and the duties we have to each other in sustaining the environment as, in the economic language upon which I will draw heavily throughout this text, a *public good.* "Environmentalism" in name enjoys a good deal of public support; mass "environmentalism" in practice is uncharted territory. Would backing for it survive in a scenario where there are great demands upon us and where we might be asked to live our lives far more simply and with the help of fewer natural resources than we consume now, directly and indirectly? Is there a moral justification for what might strike us as draconian legal and fiscal requirements upon us all on behalf of a sustainable natural environment? Is there justification for allowing others to weigh in on the value that *we* might place on it, for deeming such a value assessment one we cannot make unilaterally, like the decision whether to purchase a better cell phone?

I believe there are such justifications, but that they are neither obvious nor currently part of our common moral lexicon. These are outlined in the ensuing pages, perhaps more technically than might be comfortable for some unaccustomed to patient philosophical argument. I also take more seriously the sincerity and plausibility of many libertarian and free-market arguments than many professed environmentalists might find necessary. I will take great pains to reconstruct their case as convincingly as is possible, but if it is indeed true, as I've speculated, that such arguments may gain traction if we find that our lives will be greatly affected by the moral weight of ecology, better we should be clear sooner rather than later just what we may legitimately ask of ourselves and each other in staving off further serious damage to the planet that supports us. Indeed, the task of sorting out a fair distribution of the burdens of achieving and maintaining a sound natural environment transcends climate change and reaches to a host of situations where environmental "goods" are at stake. If a consensus on climate change emerged tomorrow, we would still have disagreement over environmental policies, because they refer to a variety of things perceived by some to be of value that cannot be readily captured and priced on a market—unpolluted air, potable water, arable land, etc.—and which cannot simply be allocated on the basis of converging supply and demand in the same way brands of toothpaste or athletic shoes might be. The notion that the environmental benefits we all share (whether we seek them out or not, as will become clear) can be priced and thus eschewed like many other things some people value and others don't enjoys significant support by influential people now, and this endorsement may well grow more extensive. My aim is to put it in perspective and to suggest that the case cannot be made on solid moral grounds.

NOTES

1. These figures were actually gathered from the Worldwatch Institute's *State of the World 2009* (Starke, 2009), particularly the articles from Flavin and Engelman (2009) and Hare (2009). As with much of the content of the volume, they are based on findings of the IPCC's Fourth Assessment Report, released in 2007.

2. Radical environmentalists are not alone in this regard. Much of radical, anti-capitalist thought has proclaimed the problems of capitalism to be largely self-engendered and self-contained. Marxists prior to or critical of the "socialist calculation" debates of the 1930s had argued that the concerns of political economy are artifacts of competitive markets which would be rendered irrelevant in a rationally planned economy, and that communism would usher in conditions of abundance under which socially desirable goods would be provided for "free." Those who took this to mean that the costs of the provision of such goods under communism would be less than the costs of their provision under markets found grounds for a reasonable debate (although "free" was thus a poor choice of words in this case). Those who took this to mean that there exist conditions of "abundance" under which productive choices *would entail no costs*—not even the opportunity costs of foregone production—and thus no need to weigh alternative allocations, met with a great deal of disappointment (see Steele, 1992:Chapter 12, as well as Nove, 1983:15–20, for an elaboration of this issue).

3. Readers already familiar with some of the area delineations within the general field of *moral philosophy* might anticipate from this description that matters of *distributive justice* will be a large part of our concerns here. This would be somewhat misleading. I will not be so much evaluating the differential impacts of environmental policy on rich and poor or the distribution of wealth *per se*. Rather, I will be suggesting (in Part II) that individuals must have some distributional "theory" of the costs and burdens of a healthy environment in mind before they can attach a "value" to those environmental goods—indeed, before they can even determine whether they or anyone else should be expected to contribute toward those goods. The "cooperative duties" to which the subtitle of this volume refers, I will argue (primarily in Part I), are more or less universally and equally distributed. This said, however, issues of *distributive justice* are difficult to avoid in any discussion, and we will touch upon them at various points.

4. See Nordhaus (2008). Bjørn Lomborg, one of the more vocal advocates of this "cautious" approach to mitigating the effects of global warming, bases much of his case on Nordhaus and the DICE model. See, e.g., Lomborg (2001, especially Chapter 24).

5. See Stern (2007). Nordhaus (2008:Chapter IX) attributes differences in Stern's modeling largely to the fact that Stern opts for a low- or even zero-discount rate on future costs and benefits—effectively weighing them more heavily in current calculations than do most economists. See Gardiner (2013:Chapter 8) for a helpful survey of the ethics of discounting and an interesting defense of the zero-discount rate approach in weighing future environmental costs, and Ackerman (2010) for a thorough survey of the different methodological assumptions between Nordhaus and Stern.

6. The philosophical patriarch of Deep Ecology, Arne Naess, took a somewhat more nuanced view on the weighing of various human and nonhuman values than did many of Deep Ecology's professed adherents (and, arguably, their contemporary progeny among "Deep-Greens"). He notes in the germinal essay which coined the term "deep ecology" that "biospheric egalitarianism" can only be respected "in principle . . . because any realistic praxis necessitates some killing, exploitation, and suppression [in and of the nonhuman world]" (Naess, 1973:95). Elsewhere he draws a distinction between the "vital needs" and "peripheral interests" attributable to both human and nonhuman nature, suggesting that such a line would allow us to assess when human intrusions into nature are acceptable and when they are not: "Humans have no right to reduce [nature's] richness and diversity except to satisfy vital needs" (Naess, 1986:14); "A vital need of the nonhuman living being A overrides a peripheral interest of the human being B" (Naess, 1984:267). Despite the similarities, Naess' "biocentric" Deep Ecology remains a far cry from the position expressed in claim (b). The "values" referred to here assume an anthropocentrism which would assert that, while the nonhuman world may be considered an object of moral concern, it cannot be said to have "needs" or "interests" independently of a valuing, human, moral agent. Whether or not this position is ultimately defensible is, as I've already suggested, beyond the scope of this book. For an overview of the Deep Ecology movement that flourished in the 1980s and arguably became the cornerstone of the contemporary radical ecology movement, see Devall (1980) as well as Devall and Sessions (1984, 1985).

7. Though the wording here is not inaccurate, it might leave one with the impression that the sorts of costs that pollution imposes are different in kind from the sorts of costs that pollution *abatement* imposes. The "marginal social costs of pollution" for many of us might invoke tangible expenses such as public health, human longevity, and the like, while my account of the "marginal social costs of pollution abatement" sounds like strictly opportunity costs of forgone production or resources not spent on other areas deemed social problems. For many contemporary economists, however, *all* costs are in some sense opportunity costs, including those costs we might naturally think of as "social ills."

8. David Ramsay Steele's point concerning recycling is especially instructive for environmentalists (though issues of environmental economics are not his principal concern). It is not the case, he suggests, that the more solid material an economy recycles, the more "waste" is reduced. The "waste" reduced by reusing and recycling materials can well be offset and even exceeded by the costs of doing so

(e.g., the opportunity costs incurred by reduced research into more efficient packaging). Indeed, "'Preventing waste' at a loss is wasteful" and "...implies that we should find some way of ensuring that actual recycling neither exceeds nor falls short of the optimum amount of recycling" (Steele, 1992:308–309).

9. See, for example, Wenz (1988).

10. We will not return to the intergenerational distributive implications of environmentalism until the Afterword. As indicated earlier, distributive justice is not the primary concern of this book, and readers interested in environmental justice might find the relative scarcity of such discussions unfortunate. However, these are issues that demand a forum in their own right and would require a separate book.

11. In the case of the United States, the informal assessment of public policy in terms of its potential costs and benefits has characterized regulation for much of this century, but the official commitment the strictures of CBA emerged in the Reagan administration by virtue of Executive Order 1229. See Kneese (1984) for an overview of CBA in environmental policy-making.

12. See Gardiner (2013:250–251) on this point.

13. One of the more common, albeit controversial, technique used to establish the market value of environmental quality is referred to as the *Contingent Valuation Method* or CVM. CVM typically employs attitude surveys to determine value according to the prescriptions of welfare economics; that is, the value of a good to an individual is just the greatest amount she is willing to expend to obtain it, or, conversely, the smallest amount she is willing to accept as compensation for foregoing it. I suggest above that "assuming risk" is more general than "paying" a cost since costs would ideally need to be assessed in regard to both their magnitude and the probability that they would accrue: hence the "risk" of having to bear them; this is the general assumption in CVM's and similar methods' application in risk assessment (see Shrader-Frechette, 1985, and Slovic, 1991). CVM as well as the larger question of environmental valuation will constitute one of the primary concerns of this book, and we will return to the issue in much greater detail in Part II.

14. "Objective" or "intrinsic" values typically refer to those which are the real properties of objects and which (a) can be said to exist *in the absence of* valuing agents, or (b) can be described *without reference to* valuing agents (See O'Neill, 1993:15–22).

15. See Edwards (1987) and Jacobs (1991) on this point.

16. This is also why environmentalist who condemn marketable emissions trading on the grounds that it "commodifies" pollution (by granting polluters the option of purchasing the "right" to do so) rather than reducing it are thus missing the point. We are not, as their objections might otherwise imply, left with as much carbon dioxide or high-sulfur emissions "as the market will bear." The EPA continues to set maximum emissions standards; it simply uses marketable incentives rather than uniform restrictions to achieve them (indeed, Hays, 1995, argues that the relatively robust legislative mandates prescribed by the 1990 Clean Air Act revisions—virtually halving the maximum total sulfur emissions allowable—vastly overshadow the "market-like" system designed to implement them). Hence advocates of "pure" markets shun tradable emissions and recommend private bargaining among those effected as the

only "true" market solution to sulfur emissions and other air pollution problems (e.g., Anderson and Leal, 1991:147).

A better critique of emissions trading might suggest that these contrived markets may impede innovation. Unlike "command and control" environmental policies, the EPA is not instructed in the course of setting marketable emissions limits to pursue new technologies that would minimize environmental damage. Tradable emissions advocates claim that financial incentives would spur technologies that reduce the cost of compliance with reduction goals, and that mandated forms of innovation are inefficient. Yet there remains in the market for emissions credits an incentive for credit-futures speculation. I can purchase, for example, year 2025 credits in 2015 with the intent of selling those credits to those firms that most need them ten years hence. In the interim, I have an (apparently perverse) incentive to keep the costs of compliance to reduction goals high so that the value of my credits in 2025 will be as valuable as possible and still provide a cheaper alternative to compliance. Hence if I pay $x per credit now, I will do everything in my power to ensure that the cost of reducing a comparable unit of sulfur emissions in 2025 is substantially more than x, lest polluting firms conclude that its simply cheaper to reduce emissions than to purchase my credits. There may not be much that I alone can do to impede innovation, but one can readily extrapolate from this example a situation where entire classes of emissions credit speculators act to keep the costs of emissions reduction higher than they might be had no such markets existed.

17. O'Neill (1993:110) draws a useful distinction between *strong* and *weak commensurability*. The former suggests that there is a single source of value for all things, while the latter is committed only to a single *measure* of value, regardless of its source(s). Weak commensurability, I am suggesting here, is sufficient for CBA. Robin Hahnel (2011) describes the situation somewhat differently. For Hahnel, practitioners of CBA are forced to resort to archaic versions of welfare economics beholden to utilitarian theories of *cardinal utility* (i.e., quantifiable measures of just how much pleasure, satisfaction, etc. one experiences from a good) in order to make comparisons across individuals—since we would need to be able to say things like "Peter will be nine units better off than he was before, while Susan will be five units worse off than she was before" (Hahnel, 2011:14). Hahnel's assertion may be on the strong side; it's conceivable that many practitioners of CBA see themselves working from a post-utilitarian *ordinalist* tradition, though Hahnel's point is merely to suggest that CBA necessarily ties itself up in implied normative claims, something I will argue shortly is true for any commitment to "efficient" policy goals. The commensurability question as well as the importance of the distinction between *cardinal* and *ordinal* assumptions in economics will be taken up once again in greater detail in Part II.

18. Nothing asserted here is intended to dismiss the obvious precautions one must take against the manipulation of "trade-offs" in environmental policy debates, nor the skepticism with which one should justifiably treat the "jobs-versus-environment" sorts of arguments harnessed by a number of corporate interests. John Bellamy Foster (1993) has detailed the problems of such approaches in the rhetoric of logging firms in the Pacific Northwest of the United States, particularly in their confrontations with environmental groups seeking to protect old-growth forests.

19. It would be a different fallacy—yet no less a fallacy—to defer to a loose form of holism and assert that such problems are inextricable, as some environmentalists are inclined to do. While it may indeed be the case that social-economic problems of poverty and the like are, at root, a result of the same institutions and practices that cause environmental problems (loosely described as *capitalism*), this doesn't rule out a relatively autonomous "trajectory" for each problem, nor does it preclude the possibility that treating one disproportionately may exacerbate the other.

20. Much of this section is derived with minor revisions from Tverdek (2004). It is used and reprinted here with the permission of The Guilford Press.

21. These figures are for 1993 and are derived from the *Statistical Abstract of the United States.*

22. See Hansel (1995). These admittedly rough values were calculated from the already rounded raw figures Hansel provides. They, like the previous ones from the *Statistical Abstract of the United States*, are selected for purposes of antedating the market volatility (primarily in the form of the tech bubble and the mortgage crisis) of the past decade, and are thus intended merely for illustrative purposes.

23. It might be more accurate to say that *three* parties are the minimum requirement for basic economic exchange: the two that actually do the exchanging, and an implied, impersonal third party representing the competitor who can offer either party a better deal. Just as it would make little sense to describe the "price" of my grain if I were socially isolated, it would make as little sense to describe one unit of my grain "priced" at two units of fish in a dyadic society: exchanges are not based on "prices" until they are made in light of available, competing alternatives. One of the most interesting treatments of this notion comes not from a classical economist but rather from the sociologist Georg Simmel, in his account of the *tertius gaudens*—literally, the third party who enjoys an advantage as the result of a quarrel, competition, etc. between the first two (see Simmel, 1950:154–169).

24. It is sometimes necessary to distinguish between the two most prominent criteria for economic efficiency, *Pareto superiority* (one state of affairs is Pareto superior to another if no individual can be made better off without making someone worse off) and its *Kaldor-Hicks* adjustment (one state of affairs is superior to another if, though some individuals may be made worse off, those made better off are advantaged to a degree which would allow them, in principle, to compensate the disadvantaged). For this discussion, I am avoiding such distinctions and referring informally to an optimal allocation of social goods such that, under any other allocation of the same productive resources (wealth, labor, etc.), the net social product would be less.

25. All of this said, it is important to also note that, however laden with moral import the term *efficiency* might be, Adam Smith's "invisible hand" does not really counsel the pursuit of self-interest on normative grounds—even if it could be said to be "efficient" in maximizing social utility. The "invisible hand," in other words, is *not* a normative prescription for unbridled self-interest, as it is often mischaracterized. In asserting that altruism is superfluous in the pursuit of the "social good," one is not necessarily asserting that self-interest is itself "good," or that it ought to be considered an end in itself. Smith—as well as much of the classical economic and rational choice traditions which succeeded him—is not offering a moral instruction to act on one's

own behalf in spite of the well-being of all. He is, rather, specifying the conditions under which he believes the former can translate into the latter. This is (regardless of the normative content in one's conception of "well-being" or "efficient") an empirical question, and to suggest that the "invisible hand" thesis endorses a selfishness that would lead to self-destructive capitalist "greed" misses the point. If Smith is to be faulted (and, as we will see later, arguments from "counterfinality" suggest he ought to be), it is because he was confused on the facts of the matter—on whether some "general interest" emerges, under particular conditions, from the pursuit of self-interest—and not simply because one may believe he expounded a morally corrupt philosophy of avarice.

Which is not to deny that some have indeed advocated avarice on behalf of the "public good" (most notably Bernard Mandeville in his *Fable of the Bees*) or to suggest that there are no normative problems with the "invisible hand" thesis. One of the more perplexing of these is its implications for utilitarianism. A common complaint leveled against utilitarians is that they are indifferent to the aims of moral agents; as long as one is acting in such a way as to maximize social utility, one's motives in doing so are irrelevant (and our nonutilitarian intuitions tell us that moral actions should be guided by moral motivations). The "invisible hand" thesis seems to endorse a hypothetical society of unenlightened utilitarians: everyone's actions effectively maximize social well-being, yet no one has the slightest inclination of doing so. Historically, utilitarians have had mixed reactions to this possibility. Henry Sidgwick, for example, seemed undisturbed by the prospect in *The Methods of Ethics* (see Cocking and Oakley, 1995:fn.3), while Ludwig von Mises, a major proponent of the Austrian School of economics and staunch defender of both free enterprise and utilitarianism, was less sanguine; he considered, according to later Austrian School devotee Murray Rothbard, "the fact that many if not most of the consequences of the market economy and society are unintended...a cause for alarm and not celebration" (Rothbard, 1993:206). Irrespective of how professed utilitarians treat this seeming paradox, it creates some disturbing self-referential problems for any defense of utilitarianism. Why bother, we might well ask, rationally persuading people of its merits when we need only dupe them into adopting behaviors that will yield the same effect? If utilitarianism is true, one should adhere to a utilitarian ethic only if, in doing so, one tends to maximize social utility. Yet, as Smith suggests, concern for social well-being may well be less effective at actually attaining that well-being. Utilitarianism thus becomes saddled with the burden of both eliciting utilitarian actions from individuals and persuading them not to *be* utilitarians.

26. This isn't to suggest that radical environmentalists see commodity production as *the* defining characteristic of capitalism (as opposed to market economies in general). Indeed, commodity production can be found in noncapitalist economies, both theoretical (market socialism) and historical (those precapitalist societies Marx described as having "simple" commodity production). Most of my comments here attribute to both market advocates and environmental critics the belief that production for exchange is a necessary but not sufficient marker of capitalism, and they would presumably wish to consider things like capital accumulation and wage labor equally essential.

There are, moreover, good reasons for considering whether the distinction between *use* and *exchange* value is itself one that makes sense only in respect to commodity production. Critics who doubt the universality of *use value* that they take Marx to be asserting in his economic writings (e.g., Sahlins, 1976, and Baudrillard, 1981) point to, among other things, anthropological evidence which suggests some cultures apply no meaningful sense of "utility" to the goods they produce, and conclude that "utility" cannot be something inherent in the good itself.

27. Outside the realm of environmentalism, similar arguments have fueled "over-productionist" claims from sources as diverse as John Kenneth Galbraith and Marxist economists Paul Baran and Paul Sweezy, all of which pin much of capitalist instability on the fact that the social surplus product must be consumed by a monumental "sales effort" to market commodities for which consumers were not aware they had a "need." It's not abundantly clear, despite these arguments by environmentalists and others, whether the proponents of capitalism have always considered endless growth inevitable, desirable, or even a simple matter of ever more product. A number of classical economists such as David Ricardo believed, for example, that capitalist growth would eventually level off into a stationary-state economy (see Wrigley, 1994). As valuable as these discussions may be in gaining a better understanding of contemporary capitalism, they arguably assume a narrow understanding of "growth." When we produce more goods, we increase the number of *options* consumers have to satisfy their wants, regardless of the degree to which they choose to satisfy them. Similarly, increased production per unit of productive resources (labor, raw materials, etc.)—the type of productive growth in question here—is as we've seen a measure of *efficiency.* If we are producing more with less, we are again maximizing the options available to individuals; should they choose not to consume more goods, they can "cash out" that increased productivity in the form of more leisure time, or time in general for which they are not compelled to labor to satisfy their needs; this is essentially what Juliet Schor describes as a distinction between *extensive* and *intensive* growth (Schor, 2010:16). Thus no one is forced to consume any more than he or she desires, nor does the surplus generated with higher productivity translate necessarily into unsold goods (however much this tendency may be exhibited, for various reasons, under capitalism). This is not to offer an argument for the inherent desirability of economic growth. The point is to suggest that the true-but-trivial claim that there are only so many microwave ovens which we can consume is insufficient in itself to establish the inherent *un*desirability of economic growth. For a good account of the economic issues involved in this question, see Beckerman (1974) as well as Baldwin (1995).

28. Stern (2015:32). The entire second chapter of Stern's most recent volume examines the "bright future" that the transition to renewable energy promises.

29. In *Plenitude*, for example, Schor suggests that the lifestyle adaptation she recommends "is available right now. It doesn't require waiting for the clean-tech paradigm to triumph" (Schor, 2010:3).

30. Despite his risk-averse position on the question of how urgently emission limits and the transition to renewables might be implemented, Bjørn Lomborg is equally skeptical of the "appropriate lifestyle" wing of the environmental movement,

accusing the IPCC going beyond its limited platform on climate change and speaking to broader issues of sustainability and equity. See Lomborg (2001:318–322).

31. See Francis I (2015). The numbers here refer to the paragraphs of the document, which are, as with all papal documents, enumerated consistently across printings.

32. It's worth noting that Stephen Gardiner's recent and quite compelling work in environmental ethics, *A Perfect Moral Storm: The Ethical Tragedy of Climate Change* (2013), considers moral duties largely to nations and other institutional actors rather than individuals.

33. A 2006 report of the Food and Agriculture Organization of the United Nations, *Livestock's Long Shadow: Environmental Issues and Options* (FAO, 2006:272), suggests that the raising of livestock contributes nine percent of carbon dioxide emitted into the atmosphere, 37 percent of methane, and 65 percent of nitrous oxide—and is thus responsible for 18 percent of the global warming effect—a greater impact than that of road transport worldwide.

34. Schor (2010:10) reports that surveys she has worked with since 2004 have indicated that as much as "80 percent of the population agreed that protecting the environment would require 'most of us to make major changes in the way we live'." It is indeed a promising figure, though Schor makes no mention of whether or how widely percentages of this magnitude are replicated elsewhere.

35. The Property and Environment Research Center (PERC) based in Bozeman, Montana, for example, harbors scholars who are in fact deeply committed to environmental quality but who maintain that free-market strategies are more effective in achieving them. Like the "techno-greens," they thus view environmentally sound technology as potentially profitable and constitutive of a thriving capitalist economy, but they eschew what they see as "artificial" subsidies for those technologies. Anderson and Leal's *Free Market Environmentalism* (1991) remains the founding monograph for this approach. Demonstrating that the assumptions these approaches make are morally confused is among the primary aims for the present work; I do not, however, wish to impugn the motives of self-described "free-market environmentalists."

Part I

WHO SHOULD BEAR THE COSTS AND BURDENS OF ENVIRONMENTALISM?

We might think it uncontroversial to claim that we are not obligated to pay for things which we don't solicit yet which are given to us by others; I don't, for example, "sell" you my watch by depositing it on your doorstep and sending a note demanding compensation. And we might further assume it to be irrelevant whether we find these things enjoyable or whether we loathe them (you might in fact find my watch delightful); the fact that we didn't ask for them seems to entail that we don't owe anyone for their provision. What if, however, the things in question were things that both you and I desired and that conditions were such that, in my efforts to secure them, they would be necessarily made available to you as well, though you had made no effort to help me acquire them on our joint behalf? Would you then be obligated to contribute toward securing these things, even if you had consented to their provision no more or less than in the case where they could be withheld from you? What if, moreover, the things in question were not simply "made available" to you but virtually inescapable to you, such that you couldn't help but "enjoy" them? Would this affect your obligation to contribute to their cost?

Consider an example. Suppose that Smithers and Burns have the opportunity to consume a good provided by the collective efforts of their community—say, a new water filtration facility for which local environmental groups are taking up a collection. Suppose further that Burns takes his chances as what is commonly referred as a "free rider." He values the environmental good in question, and would prefer having it to not having it, but he would also prefer having it and *not* making a contribution toward its provision to having it and making such a contribution. He reasons that it would be senseless to contribute toward the provision of the environmental good since withholding his contribution will make little difference as to whether the facility can be built, and since—in this particular case—the good (cleaner tap water) cannot practically be denied to him once it is provided. And suppose that Burns succeeds. A sufficient number of individuals contribute enough of their resources to fund the filtration facility, and Burns enjoys the resulting cleaner water in this community at no cost to himself.

Now since Smithers also lives in this community, he is confronted with the same options: contribute toward the new filtration facility, or withhold such a contribution. Suppose that Smithers also decides not to contribute any of his efforts toward the cause of the environmental groups and, like Burns, withholds his contribution. Unlike Burns, however, Smithers is sincerely indifferent to the prospect of having appreciably cleaner water. He has little or no regard for it, is content with the old water quality, and chooses not to "purchase" cleaner water with his contribution in much the same way he chooses not to purchase asparagus, to which he is also indifferent, if not averse. Yet whereas Smithers can easily avoid asparagus—he simply refrains from buying it at the produce market—he is more or less compelled to "enjoy" the cleaner tap water that others have provided. While Burns is enjoying a good for which he did not pay because he cannot be *excluded* from it, Smithers is subject to a good that others desired—and paid for—because he cannot readily be *excused* from it.

Is there a morally significant difference between Burns' situation and Smithers' situation? This would seem superficially so. Burns is, in some sense, getting something for nothing—something which he consumes at a rate much higher than that for which he is willing to pay (in the language of economists, he is enjoying a *consumer surplus*). While Smithers is also "getting something for nothing," it is not something which he would have otherwise preferred; even though he is not being compelled to pay for anything, the actions of others have ensured that he will, like it or not, drink cleaner water than he would have had easy access to had they not cooperated. This difference is accentuated when coercion is brought into the example. Suppose that, although voluntary contributions were sufficient to build the filtration facility, economic forecasts suggest that the maintenance costs of the facility will be more than initially expected. The same environmental activists that secured voluntary contributions for the initial project now successfully lobby the local legislature to assess a tax aimed at supporting the ongoing function of the facility. In one fell swoop, Burns is made to "pay up" for what he has been enjoying for free—an outcome that many might see as morally unobjectionable. Poor Smithers, on the other hand, is not only faced with a good he didn't care for from the start, but also for which he must now *pay* to ensure that *others* may enjoy it. This may well strike us, on the surface, as morally disturbing.

Oddly enough, it is about Burns that both environmentalists and defenders of the market have a good deal to say in examples of this sort. Most of the concern in situations like these is over "free riders" and whether coercion is necessary or permissible to prevent them. Poor Smithers—the "honest holdout," as I will be describing individuals in his position—doesn't get much attention, even though coercing *him* to pay for the provision of the environmental good seems intuitively more objectionable. These chapters in Part I are aimed primarily at reconstructing some of the arguments—often implied and unarticulated—which underlie these strong libertarian intuitions. We will see that, despite the superficial comfort we might enjoy in holding them, they are not on as solid a normative foundation as they might initially appear.

It might be tempting to counter these intuitions by claiming that Smithers doesn't know what is in his best interest, and that we cannot be faulted for providing it for him in spite of his indifference. I will argue that this would be the wrong approach to take here. The relevant question should be not whether paternalism is legitimate in this case—that is, whether we are justified in coercing Smithers' contribution *for his own good.* Rather, our concern should be whether Smithers can justify withholding his contribution to the provision of the benefit simply by pointing to the fact that he *can't escape it.* I will not be advancing any argument to the effect that it is permissible for us to exact a contribution to environmental benefits from free riders—let alone from "honest holdouts" (though I will defend the prima facie plausibility of the former claim). What I will do, however, is clarify and defend a less ambitious yet deceptively important point: an argument which justifies coercing contributions from free riders must also justify coercing contributions from honest holdouts. To put this a different way, the principal argument I will try to defend in Part I is a conditional one. *If* we are permitted to compel people to pay for certain types of goods because we can't feasibly *prevent* them from consuming those goods, we are also permitted to compel people to pay for certain goods which they cannot feasibly *avoid* consuming. We are, in other words, justified in treating the honest holdout in the same way as the free rider by the fact that he cannot be practically distinguished from the free rider on any reasonable moral grounds. Hence a rationale that would permit us to exact a contribution from one—if it is defensible—must also hold for the other. Making this claim a plausible one, however, will require a good deal of background on the nature of the goods in question, be they of an "environmental" nature or otherwise. This is where we begin.

Chapter 1

What is "Tragic" About the *Tragedy of the Commons*?

If everyone in the world who has a wall were to install an additional window, the scene outdoors would not likely become noticeably dimmer. Similarly, one doesn't receive, in any meaningful sense, *less* of a network NFL Super Bowl broadcast than one receives of a public affairs program on a local television affiliate, in spite of the fact that there are undoubtedly more viewers sharing the broadcast of the former than the latter. Economists describe this quality of certain goods as *jointness in consumption*—enjoyment by one person of a good exhibiting jointness does not appreciably diminish the supply available to everyone else.[1] These examples, moreover, present a unique case of jointly consumed public goods—one for which not only does *one person's* consumption not make an appreciable difference in the amount available to everyone else, but everyone's *aggregate* consumption has little or no impact on the overall amount available. Sunlight is in seemingly perpetual self-replenishing supply; broadcast frequencies need never be hoarded for fear of an impending "shortage." We either tap into the total or we don't, and no matter how many of us do, the same amount remains available to additional consumers (language still fails us here, since even "tapping in" implies we are siphoning off a measurable amount).[2]

Consider a somewhat different example. There is nothing the person four rows in front of you at a musical concert can do to consume *more* of the music than you can consume, and you would not likely enjoy the concert any more had she decided to stay home. Yet were we to usher an extra four hundred people into a concert hall already filled to seating capacity, this might well diminish your enjoyment of the music, as well as everyone else's. Similarly, the car next to you on the freeway is, by itself, not detracting from your consumption of the freeway. Add a few hundred cars to the same stretch of road and the story changes. Like sunlight and broadcast waves, concert hall performances and freeway rides are jointly consumed goods to the extent that another individual's consumption of them has little or no effect on how much is available for us to consume. *Unlike* sunlight and broadcast waves, both concert hall performances and freeway rides exhibit an additional characteristic that describes a category of jointly consumed goods more precisely designated "public goods"—they admit to *crowding*. While it remains true, for jointly consumed, "crowdable" goods, that consumption by *one person* does not appreciably diminish the supply available to everyone else, it need *not* be true that consumption by *n* persons does not appreciably diminish this supply. Unlike those goods for which crowding is not a concern, it is indeed possible to affect the supply of crowdable goods if sufficient numbers of individuals decide to consume it—in spite of the fact that the marginal changes in the supply of the good affected by any one person's decision to consume are, for all practical purposes, negligible and often immeasurable.

While a distinction between jointly consumed (crowdable and non-crowdable) goods and all other goods may well be useful for economic purposes, is it a morally relevant one? In the case of non-crowdable joint goods, the answer would seem to be "no." If there is no meaningful sense in which the good can be depleted, there are no prominent actions that one might take which would be perceivable as "harm," which would distinguish malicious intentions from benign ones, etc. In the case of crowdable goods, however, the answer is less definitive. On the one hand, the actions of any one individual in consuming such a good can be said to be marginally imperceptible, and our consequentialist intuitions might lead us to conclude that actions with imperceptible effects cannot be said to be morally relevant. Yet we might think that in being a member of a group which does have the collective potential to overconsume the good and thus deplete it (or at least leave an insufficient amount for others who may desire it), any one individual's actions thus become an object of moral scrutiny, especially if the good has something to do with a healthy environment—critical to the survival of everyone.[3] We might further imagine a potential conflict between what we are motivated to do by rational self-interest and what morality would require of us. Because my consumption of the good contributes to its depletion when a sufficient number of other individuals consume it—thus precluding each of us from consuming more—self-interest appears to be at odds with the interest of all. One might argue that the potential for such a situation is morally disturbing, and that this is what would constitute our normative concern about public goods. The situation is described by a number of names, but the conditions are consistent: what is true for *each* of our actions as individuals is not true for (and potentially undermined by) *all* our actions collectively. This is the conflict at work in what has become the principal allegory for problems of commonly owned, jointly consumed public goods such as environmental goods—Garrett Hardin's "Tragedy of the Commons":

> The tragedy of the commons develops in this way. Picture a pasture open to all. It is to be expected that each herdsman will try to keep as many cattle as possible on the commons. Such an arrangement may work reasonably satisfactorily for centuries because tribal wars, poaching, and disease keep the numbers of both man and beast well below the carrying capacity of the land. Finally, however, comes the fateful day of reckoning, that is, the day when the long-desired goal of social stability becomes a reality. At this point, the inherent logic of the commons remorselessly generates tragedy.
>
> As a rational being, each herdsman seeks to maximize his gain. Explicitly or implicitly, more or less consciously, he asks, "What is the utility *to me* of adding one more animal to my herd?" This utility has one negative and one positive component.
>
> 1. The positive component is a function of the increment of one animal. Since the herdsman receives all the proceeds from the sale of the additional animal, the positive utility is nearly +1.
> 2. The negative component is a function of the additional overgrazing created by one more animal. Since, however, the effects of overgrazing are shared by all the herdsmen, the negative utility for any particular decision-making herdsman is only a fraction of −1.
>
> Adding together the component partial utilities, the rational herdsman concludes that the only sensible course for him to pursue is to add another animal to his herd. And another . . . But this is the conclusion reached by each and every rational herdsman sharing the commons. Therein is the tragedy. Each man is locked into a system that compels him to increase his herd without limit—in a world that is limited. Ruin is the destination toward which all men rush, each pursuing his own best interest in a society that believes in the freedom of the commons. Freedom in a commons brings ruin to all (Hardin, 1977:20).

The benefits of one's consumption of jointly consumed, commonly owned goods, Hardin suggests, are accorded individually, while costs of consumption are distributed among everyone involved. Facing such a situation, each individual reasons that it would be worth her while to consume an additional unit of the good, since what she stands to gain in the consumption of another unit outweighs what she stands to lose personally. Since everyone is confronted with the same options, however, and since they too can be expected to act rationally on behalf of their own interests, they too will make the decision to consume an additional unit of the good.[4] The net result is that the good is depleted, and aggregate losses come to outweigh aggregate gains: what was true for *each* (i.e., it is rational to consume another unit of the good) proves to be untrue for *all.* For our purposes here, we will describe such situations with the term used by John Paul Sartre in the *Critique of Dialectical Reason* to characterize those where one's self-interested intentions are undermined in the aggregate when others act on those same intentions: *counterfinality.* I think it can be argued that in spite of initial appearances, counterfinal situations need not require that the demands of morality be at odds with what would be suggested by rational self-interest. For this reason, I think that it would be reasonable to conclude further that jointness in consumption is not the characteristic of public goods which makes their provision a moral concern.

These are the primary points I wish to make in this chapter, and we will proceed with a few arguments on their behalf shortly. Before this is done, however, we need to ward off potential confusion over the nature of counterfinalities with a brief digression into what they are *not.* Counterfinalities, as I've described them here are not *negative externalities*, despite the fact that some environmental harms are (quite accurately) described as "externalities." While both of these fall into the more general category of the *unintended consequences* of private actions and are thus often conflated,[5] negative externalities lack the particular paradoxical relationship between "rational" individual behavior and its collective "irrationalities" which characterizes *counterfinal*ities.

Negative externalities, stated most succinctly, are the "hidden costs" of production which, while being "real" economic costs, are not borne directly by the producers or consumers of the products which generate them. Thus smokers, in a microeconomic textbook example, impose negative externalities on the rest of their society insofar as the price they pay for cigarettes fails to accurately reflect the actual cost of their habit (totaling some six billion dollars annually, by some measures), given the costs of health care, lower productivity, etc., that will ultimately be deferred or *externalized* to society at large (Chandler, 1986). There is, however, no apparent conflict between individual and collective behavior here. It does not seem sensible to say that smoking is "rational" for individuals while "irrational" for society as a whole. Nor is there a "perverse" economic incentive for individuals to take up smoking at the expense of society as a whole. In strict terms of incentive structures (holding constant all other personal pleasures and costs of smoking), the fact that something is effectively subsidized does not by itself render it rational to actually *do it* as opposed to *not* do it. Smoking may be "rational" in the unlikely event that one was compelled to choose between it and some other more costly alternative, but the fact that it is less costly than it would be in some counterfactual situation where individual smokers would bear the full cost doesn't seem to provide a "rational" incentive to engage in it.

There are, admittedly, instances in which externalities appear to acquire the paradoxical nature of counterfinalities. A few important differences, however, remain between the two cases. Consider, for example, a situation often cited by environmentalists as a paradigm case of the irrationalities of externalized costs: fossil-fuel dependence. Oil, gas, and coal products, the argument goes, remain attractively cheaper to extract and refine than their cleaner and more sustainable alternatives only because the true costs of their use are not incurred by individual producers and consumers but rather *externalized* upon a tax-paying public that must ultimately compensate for the deleterious effects of global warming, regional military conflicts over resources, etc. If such costs were *internalized* into the direct extraction, refinement, and consumption of fossil fuels, it is argued, the costs of alternative energy sources would appear properly and proportionally small in comparison. Insofar as they are not, however, no individual energy producer has an incentive to pursue cleaner, sustainable forms of energy and, consequently, only "dirty," nonrenewable forms of energy are produced, again with disastrous consequences.

Despite a superficial tendency of such a situation to take on the contradictory appearance of *counterfinalities*, there are two important differences that should be underscored here. The first involves the nature of the misperception on the part of individual producers. Counterfinality, on the one hand, as Jon Elster (1985:44) points out, "is the embodiment of the fallacy of composition." What makes the actions which are seemingly rational for each individual involved (when considered in isolation) collectively *irrational* is the false belief that if it is sensible for me to do X, it is sensible for everyone to do it—that what is true of individuals is true of aggregates of individuals. The misperception involved in counterfinalities, in other words, is not over the actual costs of one's own actions but over the costs of *everyone's* actions (or, perhaps more precisely, over the *marginal* costs of one's actions given that one knows others to have the opportunity for similar actions). One may well misperceive the costs of one's own actions in the case of counterfinalities (and thus subsequently externalize them, as do the producers and consumers of fossil fuels), but this would be an entirely contingent fact of a situation in which negative externalities and counterfinalities coincide. The misperception of counterfinalities does not *require* a miscalculation of the costs of one's own actions. It requires only conditions in which the intentions of *each* individual in doing X would be undermined if *all* of them were to do X. In the case of negative externalities, on the other hand, the misperception (be it sincere or disingenuous) is indeed of the hidden costs of *one's own* actions; one simply doesn't recognize (or ignores) the fact that others are required to pick up the tab for one's choices, productive or consumptive.

The second difference between the case of pure counterfinality and one engendering negative externalities involves the nature of the deleterious effects of individual action. We describe the former case as "contradictory" precisely because actions which would have no ill-effects if only one person engaged in them (and which are thus rational for *each* individual) have "unintended consequences" in the aggregate that undermines each individual's intentions. Thus the negative outcomes emerge if and only if every (or, at least, many more than one) individual engages in the action in question. In the case of negative externalities, on the other hand, an individual producer or consumer can generate negative outcomes (in the form of externalized costs). While such outcomes may be minuscule if only one individual is involved, and are certainly exacerbated if several individuals engage in the same cost-externalizing activity, the difference is merely a quantitative one of severity. Consider, once again, the smoking example. Unlike Hardin's herders on the commons, if Herschel were the only individual in the world who smoked, and the costs of his days missed from work, medical care, etc., were not factored into the price of his cigarettes, Herschel (and a withered tobacco industry) would still be externalizing the costs of his habit upon the rest of society, however much these costs may pale in comparison to the billions that now accrue. Negative externalities can emerge from one individual or institution, or they can emerge from several; the difference is merely one of degree. More important for our present purposes, *counterfinal outcomes by their nature require the actions of more than individual.*

But just how many more than a single individual? One might rightly point out here that a large collectivity need not be involved for dilemmas of joint public goods to obtain, and that there are many instances of seeming counterfinality in which a handful of individuals can "ruin it for everyone"—that is consume the joint public good in such a way as to preclude others from enjoying it. It doesn't require that one's entire town be out for a drive for the freeway to be considered "crowded"; surely, a concert hall already filled to capacity will be uncomfortably tight long before the last of 400-person standing-room-only crowd ambles in. Even the Tragedy of the Commons allegory would suggest that somewhat less than the full roster of herders could, in adding another head to the commons, deplete it for all. It would seem that if a joint good is to admit to crowding (and, ultimately, depletion), it must be possible for it to occur at a point where less than the full population to which the good is available can consume it to a degree where there can be no further consumption of the good.[6] This fact only serves to reinforce the sense that there is a moral question at stake with crowdable, jointly consumed public goods. We can summarize the concern in this way:

> For any group (N) of individuals (i_1 to i_N), and any jointly consumed good (G), there is a subset of those individuals (k) whose aggregate consumption of the good would deplete it, making further consumption by N impossible. Each

> individual i_1 to i_N is thus faced with a choice of consuming the good in one's own interest at the risk of denying any future access to the good to everyone or refraining from consumption. Individuals i_1 to i_{k-1}, it turns out, do no harm to the larger N by depleting the good. Individual i_k, however, changes conditions such that i_{k+1} to i_N will not have consumed the good at all. Since the marginal depletion of the good for any individual i_1 to i_N is virtually imperceptible, no individual can know with certainty that she is or is not i_k. Hence any act of consumption of the good is potentially selfish and at the expense of N, and should thus be governed by a central moral authority[7] capable of regulating the supply of the good in the interest of N.[8]

Is this characterization—a mere formalization of what can be said to be transpiring in the Tragedy of the Commons scenario—an accurate telling of the moral issues at stake? I don't believe that it is. I think it can be better argued that counterfinal situations in regard to jointly consumed public goods do not entail conflicts between the self-serving intentions of any i_1 to i_N and the collective well-being of N. What is missing is neither the motivation to act selflessly nor the external moral authority to demand it, but rather the *knowledge* necessary to make a reasoned choice in deciding whether one's consumption of the good is likely to render it unavailable to all. What is "tragic," in other words, about the Tragedy of the Commons is not that it provides conditions in which it is attractive for individuals to act selfishly at the expense of others, but rather that it provides conditions which simply make it difficult to assess the impact of one's actions, and to coordinate them with the actions of others. Since the consequences of one's actions in counterfinal situations can effectively undermine one's *own* intentions, there is no necessary disparity between one's intentions and the demands of morality. This is not a novel argument. Libertarians and other defenders of the market have maintained that counterfinal-like problems are *coordination* problems, and that individuals can have similar goals for the fate of everyone as they do for themselves. While much of what remains to be said in this chapter is concordant with these arguments, I will later argue that other aspects of public goods are indeed morally relevant, and that similar libertarian arguments are inadequate in addressing these. Still, we have come upon a key point where I believe we misunderstand the nature of environmental problems and our actions that contribute to them when we apply the Tragedy of the Commons metaphor uncritically.

Consider first of all the following problem. We've seen, on the one hand, that for jointly consumed goods—even those we would consider "crowdable"—one person's consumption of the good does not appreciably affect the supply available to everyone else. We might object that this seems true of many goods privately produced and allocated on a market as well, each of which can be readily disaggregated from the total supply available. "Surely,"

I might suggest, "the hamburger I had for lunch can't possibly have had much impact on the total supply of hamburgers that are available in the world today." This may be true, but hamburgers lack a critical characteristic which seems to be implied in the conception of jointness as I've described it here (and which Samuelson, 1954, makes explicit): non-additivity in consumption. While consumption of non-joint goods can be summed to total the amount of the good available, this isn't true of jointly consumed goods. Hence, being among the "over 4 billion served" is being just that: there was a finite number *n* of burgers produced today, and my lunch was comprised of exactly 1/*n*th of them. This cannot be said of the air I breathed today, or of the news broadcast that I listened to on a radio this morning. How, then, can it be true that joint goods may be crowdable or depletable and be simultaneously true that aggregate consumption of them is not additive? How can the aggregate supply change without some change in marginal supply? This is an important concern and it points to one adjustment we must make to the notion of "non-additivity" if we are to have a coherent picture of public goods. The consumption of crowdable joint goods must, by definition, accrue in some way—that is, such goods must in *some* sense be "additive." What remains unique, however, is the fact that their additivity is not uniform across all individuals i_1 to i_N the way they are for hamburgers and the like. And if this is true, it must also be true that their consumption is not, unlike private goods, zero-sum, in that the unit consumed by any individual i is exactly the unit of the good that is deprived to all other individuals. And if *this* is true—if one individual's consumption of a "unit" of the good leaves others with less of the good than when another unit is consumed by someone else, we have no reason to believe that all individuals i_1 to i_N, as the Tragedy of the Commons would otherwise suggest, are faced with the same type of choice.

This point may be made a bit clearer if we consider an example. Imagine a small town facing a water shortage due to drought conditions and a subsequent drop in the local water table. The attendant decrease in the availability of tap water triggers two events: (1) an increase in the demand for (and, consequently, the price of) bottled, distilled water at the local market; and (2) civic plans calling for the voluntary restriction of the washing of cars. What would water consumption in this town "look" like in this situation? Figure 1.1 depicts what we might expect in the case of the private good at hand: bottled water. The vertical axis in this figure depicts the total amount of water available for retail sale in the town's grocery market, and can readily be measured in terms of, say, one-gallon bottles. The horizontal axis represents, on the other hand, individual units of consumption, which we could increment similarly in units of one-gallon bottles.

What should strike us most prominently about this figure is that the amount of bottled water available at any one point appears to be a linear function of the amount already consumed. There's nothing counterintuitive about this: for every one unit of water that is consumed, we would expect that one less unit would be available from the store. Just when this unit is purchased (i.e., at which point along the horizontal axis) would make little difference—one unit of consumption is the same marginal unit whether it is the first one consumed or the last, or at any point in between. We can see this graphically in Figure 1.1. The amount which individual B removes from the total available supply of bottled water by purchasing an individual unit beyond A's consumption of the last unit is exactly one unit. Similarly, when individual D consumes a one-gallon bottle of water, the marginal impact of her purchase over that of C amounts to a corresponding one-gallon decrease in the total supply available—the same as in the case of B's marginal impact on the total supply. Hence the area A'B'A" is equivalent to the area C'D'C". With the private good of bottled water, then, units of consumption have a fairly straightforward, linear relation to the total supply. Even if the individual consumer was incapable of inspecting the total supply available at a given point (we needn't require that he keep tabs on the store's water inventory), he can be fairly certain about the effect of his consumption on the amount of bottled water available. And assuming that the grocer keeps her shelves as well-stocked as they can be, if there were a consumer whose unit of consumption would clear the shelves, he would be abundantly aware of this fact.

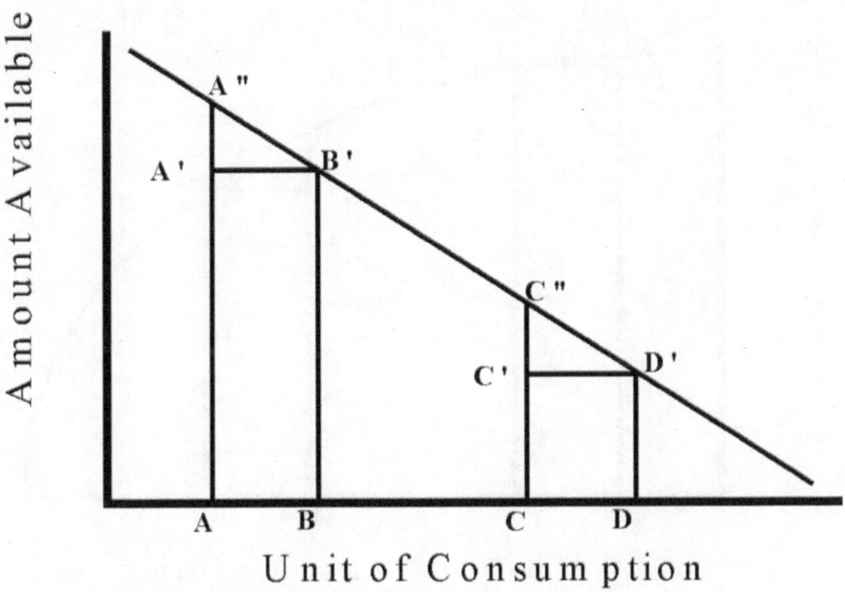

Figure 1.1 Unit of Water Consumption by Amount Available

Figure 1.2, on the other hand, represents a somewhat different situation. In this case, the water in question is made available through the local reservoir, rendered potable at a municipal treatment facility, and channeled to the taps of homes and business throughout the town, for use in drinking, cooking, cleaning, etc. What is thus at stake here is not so much the total supply of water available (since this is in constant flux, depending on natural conditions, the efficiency of the treatment facility, and a number of other factors) but rather the water *pressure* available at any one point. This we might consider a jointly consumed yet crowdable/depletable public good, insofar as *one person's* unit of consumption of tap water *may* have little impact on the available water pressure. Can we say, however, that *every* individual's unit of consumption has a marginally imperceptible impact on the total supply of water pressure available? If Figure 1.2 is an adequate depiction of this situation, this is not likely. The vertical axis in Figure 1.2 represents the supply of available water pressure which, presumably, bears some relation to the amount of water available in the reservoir, though it is not identical to it. The horizontal axis, on the other hand, again represents individual units of consumption. Since the act of consumption in our contrived example is washing one's car, we can call a unit of consumption here some constant *x* number of gallons required to do the job (assume for the sake of simplicity here that sport-utility vehicle owners are more efficient than their counterparts with compact cars and manage to clean a larger object with the same volume of water).

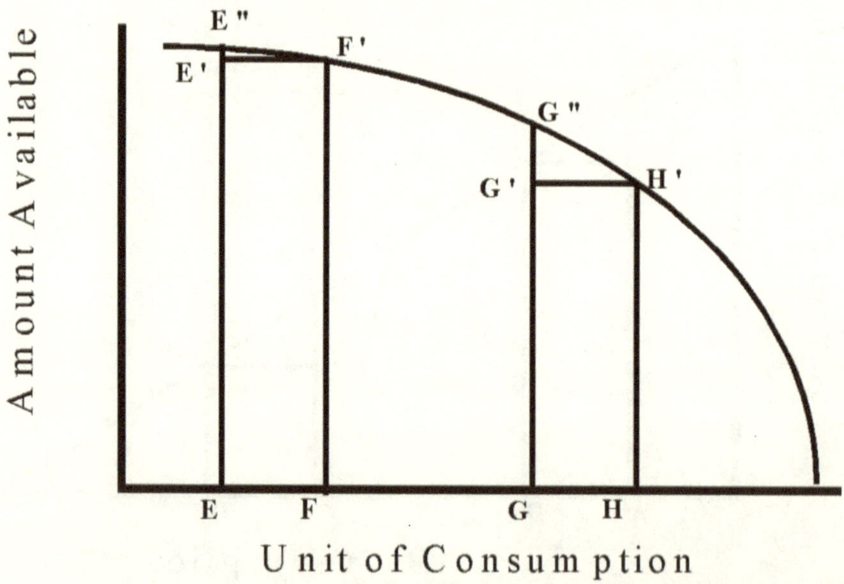

Figure 1.2 Unit of Water Pressure Consumption by Amount Available

Now suppose that in this town, all residents have been made aware of the suggested voluntary ban on washing one's car, and have been informed that this is an effort to maintain water pressure at a level which would accommodate things the local citizens' council had deemed more important, such as water for cooking, personal hygiene, and fire-fighting. Most citizens would probably agree that the ban has noble intentions, and that eating, regular showering, and dousing fires are conceivably more important to them than a clean automobile. Many of these same citizens, however, might begin to reason along the lines suggested by the Tragedy of the Commons: "Granted, it would be bad if we were all to wash our cars. How much difference would it make, however, if I alone were to sneak out tonight and wash mine? I can't imagine that the *x* gallons it takes me to do this are going to have much of an impact on the total water pressure available. Sure, others will reason in the same fashion, and may wash their cars as well (with potentially devastating effects). But if they are as clever as I am, they too will realize that others are reasoning along the same lines and will, in the interest of preserving adequate water pressure for things more aptly described as 'necessities', refrain from washing their cars. Hence I might as well sneak out tonight under the cloak of darkness and give my car a much needed hosing-down." The first of many individuals who might persuade themselves in this way to wash their cars may well be right about their suspected impact on total water pressure. Even if person E in Figure 1.2 has already washed her vehicle, the marginal impact of person F consuming the same unit of reservoir water is at best faintly measurable. What happens, however, as more and more individuals sneak out into the evening and wash their cars? It would seem that if water pressure decreases at a positively decelerating rate as a function of the unit of consumption in question (assume it does for the sake of this example), the story will be different at points further out along the horizontal axis—that is, when a substantial amount of cars have been washed, and a significant number of units have been consumed. Indeed, while the unit that H consumes beyond the last unit consumed by G is no larger than the unit F consumed beyond that of E, H's impact on the total amount of water pressure available is decidedly more pronounced, as indicated by the fact that area G'H'G" is greater than area E'F'E". Thus unlike the consumption of the private good—bottled water—where marginal units of consumption had essentially uniform impact regardless of the remaining supply of water available, marginal units of consumption of the joint public good—x gallons of water from the reservoir—have a differential impact on the total supply of water pressure available. While everyone may consume the same amount, their relative impact on the reserves of the good will be different, depending on how much has already been consumed.

The important question here regards the next few individuals after H who consider washing their cars, each of whom risk reducing water pressure in the town to virtually zero, and thus jeopardizing access to water for other essential needs. Could these individuals appeal to the fact that the impact of F's car-washing over that of E was virtually negligible to assert that the effects of their consumption too would be negligible? This wouldn't seem so, since they are washing their cars after a substantial amount of water has been consumed. Would it then follow from this that, since some marginal units of consumption have a non-negligible impact on the total supply of water pressure available, the individual i_k who consumes the critical unit which reduces water presser to zero must know who he is and must realize the impact of his actions? While we can't answer this with a definitive "yes," it seems clear that neither can we answer it with as definitive a "no" as the Tragedy of the Commons would seem to suggest. Though it would undeniably be the case that the person to consume the "last" unit of reservoir water (i.e., the unit which would reduce pressure to zero) couldn't be as certain of the impact of his actions as the person to consume the last gallon of bottled water, we would have to concede that he would have a better shot at this than F, G, or even H. He enters a different situation than do his predecessors, one in which a diminished supply has a different meaning than it does in the case of a private good consumed in discrete, uniform packages. More important, the relative risks he faces, in terms of both his own interests and those of the entire town, are much different than those of his predecessors. While he may understandably be unaware of the full implications of these differential risks, this is different from claiming that the counterfinal logic of the situation compels him, like everyone else, to consume the critical unit of the good, as the Tragedy of the Commons allegory seems to depict environmentally sensitive situations.

Hardin's "Tragedy of the Commons" scenario implies that all consumers of the good are assuming the same minimal, widely dispersed risk in consuming one unit of the good in question, and that it is only in the aggregate that the good is overconsumed. My point here is that "in the aggregate" in the previous sentence need not imply the unlikely situation in which all herders stand poised on the edge of the commons, ready to release simultaneously another head, or in which an entire town puts the concerted fatal finger to the garden hose to wash all cars at once. Joint public goods are rarely if ever consumed in one fell swoop, yet this would be required for us to draw the types of conclusions from Hardin's "Tragedy" that are typically drawn. When we consider the more applicable situations in which joint public goods are consumed over wide ranges of space and time, the conditions change in important ways. No longer is each individual locked into the same calculus of uniform gains and equally uniform, imperceptible risks. At some point, the good in question is depleted to the point where the impact of additional units of consumption on the supply available are no longer negligible. One of these units will deplete the good entirely, and this unit must, in principle, be traceable to the consumptive act of one person. The problem for this person lies not in a counterfinal logic which would have him act on his perceived rational self-interest in spite of himself. The problem, rather, is that it is difficult for him to know that *he* is the individual whose actions will undermine his own interests as well as those of everyone else involved.[9]

All of this further suggests that what is wrong in an individual's reasoning in consuming an additional unit of the joint public good is not that it is potentially "selfish" but rather that it is irrational to believe that one cannot be the critical individual i_k. What we appear to have here is a variation on the *Sorites' Problem*, a problem of the vagueness with which incremental changes affect qualitative, discrete transformations. Hampton (1987:257) and De Jasay (1989:162–163) compare it to the question of how many stones constitute a "heap" of stones; we can adapt De Jasay's reasoning to our example here in order to reach a similar conclusion: the situation is "paradoxical" only to the extent that one is committed to nonsensical, contradictory beliefs.

Here's how the line of argument works. Each potential consumer of an additional unit of reservoir water (for simplicity, let's revert to the generic designation of the good, G) approaches it with the belief that her act of consumption would not deplete the good (i.e., she cannot be i_k) given that (a) it is either already depleted by the time of her choice or it is not depleted, and (b) her individual act of consumption alone cannot be sufficient to bring it from a state of non-depletion to one of depletion. Premise (a) seems to make sense, but is it possible to maintain premise (b) and still believe that the good can be depleted? It would seem that if one were to accept the notion that G can indeed be depleted, one must also accept that there exists for any individual a finite number of units of consumption (U) which could render it such, while a similar number of units (-U) could, in theory if not in practice, "restore" it to its pre-deplete state. If we were to ask our potential consumer of G whether the *next* individual to come along could be the critical one, or the next one, or the one after that, we would be able to point out the potential inconsistency here: she cannot discount an infinite number of "next individuals" and still hold that G can somehow be depleted. And in accepting this, she must also accept that for any unit of consumption of G between -U and U, it must have a definite probability of being the depleting act (and thus the consumer of G to whom it is attributable a definite probability of being i_k). Otherwise, she must believe that there is some sort of discontinuity in the probability distribution between -U and U, one in which some magical "leap" into depletion takes place irrespective of the probabilities leading up to this point, and those which follow it. Without some reasonable account of how this might work, her only option is to concede that some definite unit of consumption between U and -U is the critical one which will deplete the good, and that the fact that we can assign only probabilities to this act does not render it nonexistent, or the phenomenon of the depletion of a joint public good an "emergent" one that derives from indefinite collective acts. And without the ability to situate her act of consumption in some probability distribution "black hole," she must concede that she too has a definite probability of being i_k, even if she does not (or cannot) know what that probability is.

We are in dense terrain; I should recount what has been suggested in this chapter. The consumption of joint public goods—and environment quality must surely count as such a good—is typically thought to take on the characteristics of counterfinality, such that while it is rational for each individual to consume a marginal unit of the good, the aggregate effects of this consumption may effectively deplete the good and render it unavailable to all. Some people take this to mean that individual acts of consumption are potentially selfish and without regard to the interests of others, since their effects, while negligible in themselves, contribute to the depletion of the good (or the failure to provide it in the first place) for one's own sake. Those who understand the situation in this way may further argue that it is thus permissible for an external moral authority to regulate individual acts of consumption on behalf of the "common well-being," so that the good remains available to all. I have argued that this misconstrues how joint public goods are consumed. While it may be difficult to discern incontrovertibly that one's act of consumption of a joint public good is the one which would render it depleted, it is reasonable to argue that (a) the way in which joint goods are marginally depleted would provide at least some information to this effect; and (b) the fact that one cannot know with certainty that one is i_k does not entail that no individual can indeed be i_k. I think we should conclude from this that the "problem" of joint public goods is not one in which potentially selfish consumers confront the requirements of morality, but rather one in which individuals with limited information must coordinate their actions and assess whether those actions are those which could render the good depleted and any further amount unavailable to others *as well to themselves.* The Tragedy of the Commons account of counterfinality thus distracts us on two counts: it mistakes practical difficulties in the coordination of knowledge for *absolute* epistemic limits,[10] and it suggests a conflict between self-interest and the interests of others that doesn't survive under scrutiny. The Tragedy of the Commons allegory, is not alone in this regard, as we will see in the next chapter.

NOTES

1. The idea of such goods—*public goods*—was discussed among European economists as early as the nineteenth century, but its characterization in terms of *jointness* and its counterpart to be discussed shortly—*non-excludability*—owes its

existence to Anglo-American economists since the 1950s. Paul Samuelson's "The Pure Theory of Public Expenditure" (1954) is widely considered germinal in this area and launched a flurry of discussion throughout the 1950s and 1960s (e.g., Samuelson, 1955, 1958; Margolis, 1955; Strotz, 1958), though Samuelson was himself taking up work initiated by R. H. Musgrave in the 1930s (see Head, 1962, on this). Mancur Olson's *The Logic of Collective Action* (1965) established a psychological foundation for what would come to be viewed as "free rider" problems of public goods, later to be clarified and extended by Russell Hardin's *Collective Action* (1982). These works constitute a fraction of what should rightly be considered the canon of public goods theory, and while my intent here is to clarify a minor point concerning the relative normative import of jointness and non-excludability, the thrust of the analytic distinction described is owed entirely to this literature. A few additional, more recent works are also helpful in navigating this economic distinction: Anthony De Jasay (1989), Jeffrey Rogers Hummel (1990), and Garret Cullity (1995). While each, like his predecessors, raises his own substantive arguments, the expository clarity in their accounts of the nature of public goods (especially via the extensively annotated footnotes Hummel provides) should mark each as required reading in this area.

2. Hardin (1982:17) draws a useful distinction between goods exhibiting jointness and those simply provided in *superabundance.* While the two categories may overlap (and probably do quite often), the bountiful supply of a good need not entail its jointness in this sense. Water, for example, is often erroneously thought to be universally in joint supply, much like sunlight or radio waves. As any inhabitant of the arid regions of the world can attest, one person's consumption of a gallon of fresh water affects the amount available to others no less than one person's consumption of a pizza means there is one less pizza in the world available to everyone else (we will see in a later example that the ability to deplete water supplies renders it a *crowdable* joint good). Hardin notes that the peculiar nature of the Lockean Proviso might be attributable to the superabundance that once characterized arable land in the world; we are left wondering today just how one is supposed to mix one's labor with a plot of land, claim it as one's own, and still leave "as much and as good" available to everyone else. See Schmidtz (1991:17–20) on the question of whether there could ever be an original appropriation of property that could literally meet strictures of the Lockean Proviso.

3. Derek Parfit, in his *Reasons and Persons* (1984:Chapter 3), considers ignoring the imperceptible effects of individual actions one of the "Five Mistakes in Moral Mathematics." A brief example along the lines of those discussed by Parfit should illustrate his point. Suppose we have two torture chambers, A and B. In chamber A reside 100 torturers and 100 victims. Each torturer gives a full turn of a thumbscrew attached to a particular victim, causing the latter a great deal of pain. In chamber B also reside 100 torturers and 100 victims. In B, however, each torturer gives 1/100 of a turn on the thumbscrews of each of the 100 victims. A 1/100 turn on any victim causes no perceptible increase in pain, so we can say of each torturer in B that he does not directly bring about perceivable pain to any particular victim. Yet the situation in chamber B should probably strike us as no less disturbing than the one in chamber A. We might thus concur with Parfit that we shouldn't ignore as morally irrelevant the

imperceptible effects of our actions, since they may be actions taken as part of a group. This might appear to conflict with what I will later argue in this chapter. Suffice it so say for now that claiming there is no morally relevant difference between the actions transpiring in chamber A and those in chamber B is not to confirm just what is morally relevant in each. Given that the actions in question in such "*n* torturers" examples are ones where the intention is that *someone* be harmed as a result (regardless of whether any individual torturer is culpable in chamber B), I think Parfit's concern in this "mistake" is somewhat different from the actions that are in question here—those which involve the consumption of goods the *indirect* consequences of which, for all we know, *might* bring about harm.

4. One might notice that the Tragedy of the Commons allegory as Hardin lays it out itself appears to presuppose markets and production for exchange. Were each herder producing strictly for use, the commons need not be at risk (or it would be at risk only for "natural" reasons such as potential drought or flooding); each herder would have an incentive to graze only the number of livestock necessary for personal consumption, and there would be no need to graze more than that, since a "surplus" of livestock products would have no particular meaning or value. Only if there are already institutionalized commercial outlets for any surplus in livestock products would we expect each herder to add heads of cattle to the commons continuously. While this may not be true of all environmental situations Hardin is trying to capture with the allegory, it is ironic that the chaos we are to believe is characteristic of communal land ownership from his example is only compelling if we assume that private, marketable goods are being reaped from it. I thank Richard Hudelson for raising this point.

5. See, for example, Grundman (1991).

6. Joseph McCarney (1989:136–140) argues that Elster's description of counterfinalities as *logical* contradictions emerging from the fallacy of composition (i.e., the derivation from the claim "For any *x* it is possible that *x* is F" the further claim "It is possible that *all x* are F") is overstated. Thus, in Elster's example, while it may be true of each individual in a full auditorium that one would get a better view of the podium if one were the only person to stand up, it is impossible for *any* plurality of individuals to be the "only one to stand up," and not merely for *everyone* to be the "only one" to do so. McCarney concludes, rightly so it seems, that it is not the fallacy of composition that is at work in counterfinalities involving social situations where the actions of substantially less than *n* individuals can undermine the intentions of each of them.

7. Elinor Ostrom (1990) has argued that the Tragedy of the Commons (much as I've described Hardin's allegory here) paints a false choice between chaos and the imposition of a centralized moral authority. She outlines both the theoretical rationale and the accumulated empirical evidence from various sites around the globe which suggest that self-organized self-governance among appropriators of "common-pool resources" presents a viable alternative to state control of such resources. To the extent that such "self-organized" systems must ultimately rely on coercion to have an effect upon the actions of those who would violate the mutual agreement (even Ostrom imagines rules which are "sanctioned using graduated punishments"; see 1990:186), there doesn't seem to be any substantive moral difference between the resource

regulation described here and the resource regulation Ostrom envisions under "self-governance." Given, as I will argue shortly, that counterfinal problems such as those captured by the Tragedy of the Commons are actually coordination problems resulting from epistemological difficulties and limited information, Ostrom's recommendations—most of which assume an abundance of shared information among resource appropriators—appear to be quite consistent with the conclusions I will be drawing in this chapter.

8. Formal notation in descriptions of public goods dilemmas varies somewhat, but I have maintained the conventional delineation here between the entire group in question (N) and the critical subset (k) which can somehow make a difference in the consumption of the goods. We will see later that this same convention is used in describing problems in the *provision* of public goods, where (N) represents the population enjoying non-excludable access to the good and (k), the minimal number of individuals whose contributions are required to *provide* (as opposed to *deplete*) it.

9. To claim that the Tragedy of the Commons loses its allegorical bite when consumption is considered over time should not imply that it is simply a temporal problem rather than a spatial one, or that the real concern is thus intergenerational selfishness, as O'Neill (1993) suggests. O'Neill argues that the herders in Hardin's fable, insofar as they are rational self-maximizers, "will add to their herd and graze the land to its limits within that generation; the benefit accrues to themselves, while the loss is shared by all successive generations" (O'Neill, 1993:39). Successive generations, he argues, thus "occupy a temporal 'commons'," as real as the spatial commons Hardin had described. It is conceivable, however, that what I am arguing here on behalf of the interpersonal assessments of individual consumers can apply to the inter-temporal case of successive generations. If it would be irrational for individual i_k to consume the fateful unit of a joint public good which would render it unavailable to all—including himself—it would be irrational for any generation of individuals to deplete the good and thus deprive themselves future access to it no less than later generations. For any particular generation to, as O'Neill suggests figuratively, "graze the land to its limits within that generation" (O'Neill, 1993:39) is hardly a good idea, and given the current discussion, the counterfinal incentives would seem to be no more compelling than in the case of individual consumers within a generation. Again, the more important question would be an epistemological one of how any one generation can be sure that it is the one which stands to deplete the good; while this generation cannot, as with individual consumers, know this with the certainty they might have with private, marketable goods, the nature of many joint goods as depicted in here would suggest that they may have some idea of the consequences of their actions.

10. This is essentially the point where I depart from Carl Cranor's (2010) otherwise lucid analysis of this area. Cranor comes at the matter from the other end: moral duties. He contrasts what are known as Direct Harm Duties (DHDs) with collective duties and suggests that much of the distinction boils down to "the epistemology of moral wrongness" (Cranor, 2010:166). DHDs have the epistemic benefit of a visible correspondence between the acts which violate them and the harm that results. If, for example, we prohibit poking people in the eye with a stick, a clear DHD, my actions which violate that norm (poking someone in the eye) correspond to a visible

harm (an injured eye). Collective duties, on the other hand, show no such direct correspondence to particular instances of harm. My violation of a duty not to emit more than my share of carbon emissions, for example, may not result in an obvious harm to the atmosphere. While Cranor sees this as a categorical difference between the two types of harms and thus the two types of duties to avoid them, what we've seen in this chapter suggests that the problem is not an absolute epistemic limit (since not all violations of the collective duty are equal, given changing circumstances) but rather the *practical* impossibility of any one person's knowing with certainty whether or not his or her action in consuming (or failing to preserve) an environmental good is the decisive one that "tips the scales."

Chapter 2

Are Environmental Problems "Prisoners' Dilemmas"?

Much of what I've argued in the first chapter can be expressed in a different but related way using another common set of tools that economists and political philosophers employ to understand how individual behavior translates into social phenomena: game theory. As was the case of the Tragedy of the Commons allegory in the first chapter, however, my aim here is not to provide a comprehensive overview of the tool itself; this has been done quite elegantly by others (e.g., Hargreaves Heap and Varoufakis, 1995). Rather, I will confine my comments to a very particular situation. Some people believe that our joint decisions over whether to consume another unit of a public good constitute what is known in game theory as a *prisoners' dilemma*. There is much we could say about this proposition and how environmentalists and economists have addressed it, but I will say only the following: (a) prisoners' dilemmas constitute only one game-theoretic example of counterfinality, one in which the mutual defection of each players is preferable to being a "sucker" *and* in which the net social benefit of mutual cooperation is greater than that of mutual defection; (b) consumption decisions about joint public goods more closely approximate game situations in which these preferences are reversed—that is, in which mutual defection is relatively catastrophic and less desirable than being a "sucker." This, we will see, is really just a translation into the technical but precise language of game theory what I argued in the first chapter. As with the allegory of the Tragedy of the Commons, game-theoretical language is what many people who take an economic approach to environmentalism use to understand the problem of ecological harm, so we have a bit of a duty to speak to them in their language, however complicated it may seem, if we are to be critical of their conclusions. Let's unpack these claims, then, to see just what it means, and why it may be considered essentially the same point outlined earlier.

Prisoners' dilemmas can take on a number of conditions in modeling real social interactions, but the essential structure is much like that seen in Figure 2.1. One player confronts another player, and each has the option of cooperating on a particular venture or defecting; neither knows which option the other player will take, and neither has the ability to signal his own intentions.[1] The cells of the 2x2 matrix represent the joint options of the two players, and contain the relative payoffs for each particular outcome, with the payoff for the row player preceding the comma, and the payoff for the column player following it. The "payoffs" as they are arranged here can represent anything of value to the players; what is important to the prisoners' dilemma or to any game-theoretic representation of a joint decision is not the substance of what "players" are "competing" for but the relative desirability of the possible outcomes to each. Hence, the integers included in this example represent not known quantities of any particular good, but rather the ordinal ranking in value of each outcome to the respective players.

Given this, we can see that the overall ranking of the four possible outcomes for Player One (in the rows) would be C>A>D>B, since having 2 units of the valued good is better than having 1, which is better than losing 1, which is, in turn, better than losing 2. Similarly, the overall preference ranking for Player Two (in the columns) would be B>A>D>C, since 2>1>–1>–2. These are *nonstrategic* preference rankings insofar as they order each player's preferred outcomes irrespective of what they believe the other player is likely to do (indeed, irrespective of whether each can assume the other to be rational). In general, we can say that (i) cooperation between the two players has a particular payoff, (ii) defection by either player while the other extends a hand at cooperation yields the highest payoff for the "free rider" (this is a much

		Player Two	
		cooperate	defect
Player One	cooperate	1,1 (A)	-2,2 (B)
	defect	2,-2 (C)	-1,-1 (D)

Figure 2.1 Standard Prisoner's Dilemma

looser notion of "free rider" than is associated with the public goods problem, as we'll see; I'll use it here out of convention and for lack of a better term) and the lowest payoff for the cooperative "sucker," and (iii) mutual defection leaves both with less than they would have had they cooperated, but more than they would have had they been the lone "sucker."

Now we might assume that the structure of the game would suggest that what would be rational for Player One to choose—cooperate with Player Two or defect—would depend on what Player Two opts to do, and vice versa. Indeed, game theory is designed precisely to codify *strategic rationality*: how we decide to do things in our own interest given that others do indeed have similar opportunities and options in maximizing their own interests. What characterizes the prisoners' dilemma as a uniquely interesting game, however, is the fact that strategic rationality renders Player Two's choice irrelevant to Player One, and vice versa. In the classic prisoners' dilemma game as seen in Figure 2.1, defection is a *dominant strategy* for each player—that is, it is an option which is preferable regardless of what the other player chooses to do. Whether Player Two chooses to cooperate or defect, in other words, Player One would be better off in defecting. In the former scenario (Player Two cooperates), Player One gains 2 units of the valued good by defecting, compared to only 1 unit gained by cooperating. Similarly, should Player Two decide to defect, Player One loses only 1 unit of the valued good by himself defecting, compared to the 2 units he would lose by being a "sucker" and cooperating. The same relative values obtain when we hold Player One's choice constant and assess the relative payoffs to Player Two. In a prisoners' dilemma, *mutual defection is a dominant strategy insofar as it is rational for each player to choose it regardless of what the other player chooses.* We can expect that each player, in realizing this (assuming each is rational) will opt to defect.

Why is the classic prisoners' dilemma of interest to game theorists or anyone concerned about the supply of joint public goods, the protection of the environment, or any practical dilemmas in the real world? It is because, it seems, we have a paradox: while defection is rational for each player in assessing his own utility, it is a *Pareto inferior* outcome in the aggregate. At least one player (both, actually) could be made better off and no player made worse off by opting for mutual cooperation (cell A), in spite of the fact that rational self-interest would steer each intractably toward mutual defection (cell D). We could say to each player: "Look, you'll both be better off if you simply choose to cooperate," but it would be fundamentally irrational for either of them to heed our advice; since Player One can't know what Player Two will choose, and vice versa, he is simply better off taking the route which genuinely believes will maximize his self-interest. And as we can see, the relative payoffs in a prisoners' dilemma situation are inevitably going to fall in favor of defection for both players. Yet the equilibrium point that is reached by each player exercising his own rationality has less social utility than one in which both players chose the opposing option—it leaves each player worse off than he would be had both cooperated. Indeed, if we were to rank the *aggregate* utility of each outcome (i.e., the net social utility garnered by summing the individual values in the cells), we would see that mutual cooperation yields the highest aggregate value (2 in this example), the two free rider/sucker outcomes off of the main diagonal yield the second highest (i.e., the 2 and –2 in each case cancel out to 0), and mutual defection the lowest aggregate value (–2). Hence an impartial third party, knowing the preferences of each participant and concerned strictly about overall social utility, might rank the four possible outcomes quite differently from either player: A>B=C>D. This is the essential problem of the prisoners' dilemma. Given this, it should become clear why one might suggest that the participants should mutually and voluntarily submit to an external moral authority—one with the power to coerce individuals to act in spite of their personal, rational assessments rather than in accordance with them. Not only is this in the interest of everyone else, it is in the interest of the individual being induced to cooperate.[2]

There are a number of important concerns we might have about this conclusion, but I think the most important objection that can be raised here is not that it is fundamentally wrong (it may well be), or that, right as it may be, it captures no social choice problems in the real world, or even that problems concerning the consumption of joint public goods are too important and nuanced to be described in the form of heuristic "games" (though many environmentalists are prone to raise the latter claim). The important objection to raise is that most situations where individuals confront each other in deciding how to consume joint public goods do not really approximate a prisoners' dilemma. While the prisoners' dilemma situation requires, as we have just seen, that mutual defection (cell D) be preferable to each player over being a "sucker" (cells B and C)—since losing 1 unit of value is preferable to losing 2—mutual defection in the case of jointly consumed public goods (i.e., with each player consuming a share of the good more likely to deprive its future availability to anyone) is likely to be less preferable for each than to being the occasional "sucker." Again, let's see what this might mean.

Game-theoretic strategies, of which the prisoners' dilemma is just one example, need not be one-time decisions among two and only two individuals. Such games can be *iterated* such that each decision is one of any number that will be made jointly with the other "player," and each "player" can range from a single individual to any number of individuals—hence the label *N-person games* (see, e.g., Russel Hardin, 1982:25). Consumption decisions over joint public goods might readily be characterized as iterated games, since we make a decision about whether to consume such goods quite often, and each decision must consider not only our immediate preferences but also our preferences over the long haul. They might also be readily described as N-person games, since everyone to whom the public good is available is making a similar decision. It might be clearer for our purposes, however, to adopt only the former assumption and not the latter. Since, as we've seen above, consumption units are arrayed over some dimension of space and time, it's not really the case that one confronts *everyone* simultaneously in making a decision about whether to consume a particular unit of a joint public good. Rather, if it is to be a decision about whether to consume a full unit which would engender crowding/depletion or to ration one's unit of consumption in order to avert these consequences, it must, in theory at least, be a decision made jointly with *one other person* whose consumption decision faces the same prospects (recall that, as I've suggested above, if a public good is crowdable or depletable, one person's unit if consumption must, in theory, be sufficient to bring the good to this state—regardless of the number of other potential consumers of the good). Figure 2.2 illustrates this temporal dimension and highlights the critical individuals as they are arrayed upon it, retaining the same notation that we have been employing throughout this discussion.

Thus, rather than translate this situation into a game matrix where Player One confronts *N*—the rest of the population to which the joint public good is available, it might be more appropriate to imagine a matrix which, as far as any player is aware, would be the pivotal iteration of the "game"—one in

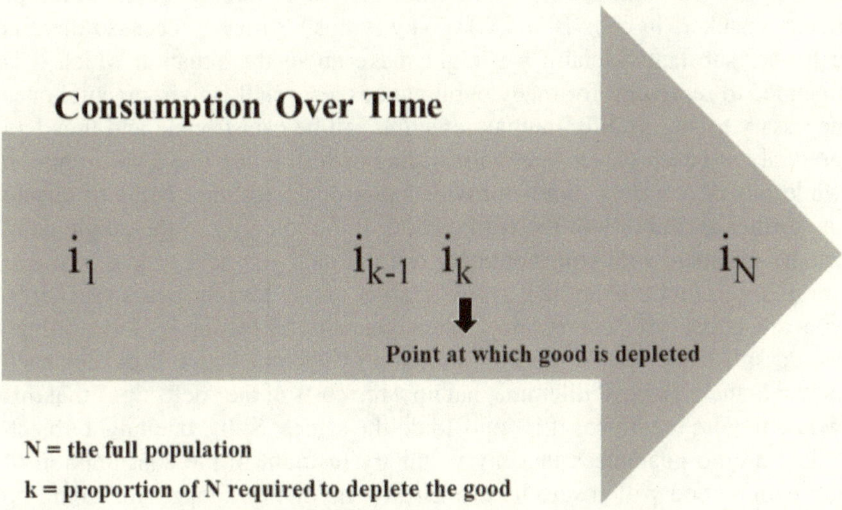

Figure 2.2 Depletion of a NonRenewable Good

		$i_{k-?}$	
		cooperate	defect
You	cooperate	1,1 (A)	-2,2 (B)
	defect	2,-2 (C)	-3,-3 (D)

Figure 2.3 Joint Public Goods Game

which each player must assume that k-2 individuals have already consumed the good and in which each player vies for the position of i_{k-1} against an opponent whose mutual defection would effectively render one i_k. Figure 2.3 pits you against this unknown individual, whom we'll label "$i_{k-?}$" to designate her ambiguous status.

Note that there is only one prominent change in the relative payoffs between the classic prisoners' dilemma situation seen above and this one: cell D—mutual defection—now carries a relative value of –3 for both you and $i_{k-?}$, a value that each of you would assess as inferior to the value of being a sucker (in cells B and C).[3] Why is this? Primarily because there is a further substantive claim we might make about the situation which it is intended to represent: for many public goods, especially environmental ones necessary to sustain life, mutual defection can be catastrophic and thus less preferable to each player than withholding or restraining one's consumption while others consume more—provided that one's restraint helps to ensure the further availability of the public good. If the integrity of the public good can be sustained with your voluntary restraint and you have at least some to enjoy in the future (even if $i_{k-?}$ took a larger share, leaving you a "sucker"), you are better off "cooperating" than you are "defecting." "Free riding" would still leave you in the best situation of all, much like it did for each player in the prisoners' dilemma, but now the costs of the "defection" that this would require are higher if $i_{k-?}$ opts to do the same. "Still," one might object, "there are no guarantees that my voluntary restraint in the consumption of the public good will ensure its availability; my act of cooperation could be met with defection by yet another player, such that the unit consumed by $i_{k-?}$

is enough to deplete the good to the point where it is virtually unavailable to myself and everyone else or, at least, what is available is vastly inferior in quality. Would this not be worse than the case of mutual defection, where at least 'I got mine'?" Not necessarily, and this is especially evident in the case of environmental public goods. Even if the good is not sustained, it is rarely meaningful to say we are better off experiencing environmental catastrophe *and* having consumed a larger share of the good than experiencing catastrophe having taken a smaller, cooperative share. Should the planet be consumed by maladies stemming from climate change, it would not likely prove much of a consolation to anyone to have squeezed a few extra miles out of his car rather than walked. To say that the depletion of a vital public good *and* having consumed a larger share of it before it was lost is marginally better than the depletion of a vital public good while having restrained one's consumption, while true in the abstract, undermines the relevance and utility of game theory for understanding such social choices. Moreover, since this particular "competition" between you and $i_{k-?}$ is only one of several iterations, it remains sensible to say that "getting yours now," if it amounts to rendering the public good unavailable to you and to everyone else in the future, is insufficient compensation to make mutual defection (cell D) somewhat more attractive than being a "sucker" (cell C) in this game. Hence the relative discrepancy in the payoffs to the "sucker" (–2) and the mutual defectors (–3) seem fitting for such joint public goods.[4]

Given the adjustments to just this one cell, and given this substantive account of the relative value both you and $i_{k-?}$ might assign to the four possible outcomes, we can see that the pure, nonstrategic preference rankings for both of you now differ from those of players One and Two in the classic prisoners' dilemma. A preference for cell A over cell D—mutual cooperation over mutual defection—was the only shared preference ranking between Player One and Player Two in the former example, and these options provided the second- and third-ranked preferences for each:

Player One: C>**A**>**D**>B
Player Two: B>**A**>**D**>C

In the current case—let's call it the *joint public goods* game for lack of a better label[5]—both you and $i_{k-?}$ share a similar preference for mutual cooperation (cell A) over mutual defection (cell D), but your preferences also share an additional characteristic not true of the classic prisoners' dilemma: *both of you prefer mutual defection least of all*:

You: **C>A>B>D**
$i_{k-?}$: **B>A>C>D**

Among the important implications of this is the fact that there is no dominant strategy in this joint public goods game. In the classic prisoners' dilemma, we saw that, regardless of what Player Two opted to do (cooperate or defect), it was better for Player One to defect; his relative payoff in defecting was superior to cooperating no matter what Player Two did. In terms of navigating the matrix in Figure 2.1, this means that the payoff for him in the bottom row is always higher than the payoff in the top row within the same column. The same is true, of course, in switching to Player Two's perspective: he too is better off defecting whether Player One chooses to cooperate or to defect; the relative payoff for him in the second column is always better than its counterpart in the first column. This is not true, however, for our joint public goods game in Figure 2.3. When $i_{k-?}$ cooperates, you are better off defecting and "free riding" on his benevolence—hence a payoff of 2 rather than 1. Yet when he defects, your mutual defection would spell disaster, and you would be better off cooperating, thus averting this—hence a payoff of –2 rather than –3. The same relative values hold for $i_{k-?}$'s decision. Where the prisoners' dilemma rendered it rational to defect regardless of what the other player did, the joint public goods game can offer no such advice; no longer does it make sense to stick to a single strategy comforted by the knowledge that, no matter what the other person does, you are maximizing utility for yourself. What it would be rational for you to do while on the brink of depleting the joint public good depends entirely on the actions of the most important person to you at that time—the other individual $i_{k-?}$ whose mutual defection is capable of rendering you i_k, thus depleting the good.

Equally important here is the fact that the paradoxical nature of the classic prisoners' dilemma is not evident in the joint public goods game: since there is no "best" strategy for either player, there exists no equilibrium point toward which both you and $i_{k-?}$ can be expected to gravitate and which is, in the aggregate, Pareto inferior. Indeed, in a joint public goods game such as this, there is no equilibrium point *simpliciter*. Hence there is no longer a question of a discrepancy between what we might expect rational individuals to do and the best collective outcome. Rather, the joint public goods game presents a question of how the players are to coordinate their decisions to maximize their own payoffs and yet avoid mutual defection. In the case of the prisoners' dilemma, as we saw, each player would benefit by acting in spite of what rational self-interest would suggest, given that their strategic assessments of the situation "locks" them into a course whose payoff is inferior for both. In the case of the joint public goods game, there is no inherent discrepancy between the self-interest of each player and the collective interest of all of them; while each might well gain the most by "free riding" on the other's cooperation, each is also bound to lose the most should the other player decide to act similarly. And unlike the conditions provided by the prisoners' dilemma, you have no a priori reason to believe that $i_{k-?}$, if rational, will act in any particular way. Thus rather than a discrepancy between self-interest and the collective good which is at work in the prisoners' dilemma, and which makes relevant (if not morally defensible) an external normative authority, the joint public goods game is plagued largely by insufficient information. It is *because* and not *in spite of* the fact that we can't know what choice $i_{k-?}$ will make that makes it problematic. An external, omniscient authority in this case might be able to share impartial knowledge of the situation (e.g., whether this is the critical iteration of the game which could deplete the joint public good; what contingencies might steer $i_{k-?}$ in toward one option rather than another). But the counsel of such an authority for both parties to cooperate would not carry any meaningful normative import; it is not our "selfishness" that is potentially damaging to our collective good in the joint public goods game, it is our ignorance.

Why, then, are these two types of games often conflated, and counterfinal problems in the consumption of joint public goods—described allegorically in the Tragedy of the Commons—portrayed as a (typically "N-person, iterated") prisoners' dilemma? Note that if we consider only the aggregate preference ranking for the four options in each case, the two games are indistinguishable. Just as A>B=C>D is true when we sum the cell values in the case of the prisoners' dilemma, it is true in the case of the joint public goods game. It is conceivable that, since mutual cooperation is collectively preferable to all other outcomes in both game situations, and since there is at least some counterfinal incentive for defection (being a "free rider" yields the highest payoff in both cases), we might be prone to lump the two strategies together. As we have seen, however, these aggregate preference rankings belie important differences in the individual strategies open to each player across the two games. In the case of the prisoners' dilemma, the fact that mutual cooperation is collectively preferable to mutual defection is meaningful in a special sense: it is precisely the outcome we would *not* anticipate if both players acted rationally. In the case of the joint public goods game, this is not true. Mutual cooperation remains collectively preferable to mutual defection (even more so, given the inflated costs of the latter to each player), but cooperation is now no longer individually irrational, insofar as its prudence as a strategy for either player does indeed depend on what the other player chooses. There is nothing in the structure of the game or the appeal of the relative payoffs which would lead us to expect each player to either consistently cooperate or consistently defect; what each would do would depend on conditions and information external to the structure of the game itself.

Thus the prisoners' dilemma is only one game-theoretic instance of counterfinality—one where mutual defection is preferred by both parties to being a "sucker"; consumption decisions over joint goods more often involve a different set of conditions in which being a "sucker" is preferable to mutual defection. While the former yield Pareto inferior outcomes (since defection is rational regardless of the other's actions) which can be avoided only by prior mutual, morally binding agreements to cooperate, the latter involves coordination problems: both parties would prefer to avoid mutual defection but have insufficient information on how to do so and still maximize their own self-interest. This is decidedly different from a situation in which self-interest will compel them to act in spite of the interests of others and of themselves.

Perhaps more important, if these game-theoretic models are to be at all useful in mapping our real-life decisions that affect the natural environment, is the fact that one can never know whether one is involved in a prisoners' dilemma situation or a joint public goods game. This is just another way of stating what we've already seen in describing the nonlinear effects of consuming a public good. In the same way one cannot be certain whether one's act—be it consumption, pollution, or whatever—is the one that "tips the scales" and brings about severe environmental harm (i.e., we can never be practically certain that we are in fact i_k), one cannot know for certain that one's "opponent" in the two-person game model is i_{k-1}. There is no difference here in what is being asserted, and this is why, I believe, it is unhelpful to describe the Tragedy of the Commons and real-life situations where individual actions can have grave environmental impacts as merely *N-Person, iterated prisoners' dilemmas*. If the conditions for every "player" were the same at every iteration, this might be an accurate description. Those conditions are not, however, and at some point k-1—the verge of depletion of the environmental "good," whatever it might be—what may have been a genuine prisoner's dilemma where mutual defection is preferential to being a sucker becomes a joint public goods game where mutual defection could be devastating and leave both "players" worse off than they would have been as "suckers." The problem is thus not a moral one as such; it is an epistemic one: one can never be quite certain that one is i_k, or (which amounts to the same thing) that one is engaged in a competitive game with i_{k-1}. Pure self-interest might ensure that we not "tip the scales" toward environmental catastrophe if we were put in such a position; the problem is that no one can ever really know whether he or she is in such a position. If the Tragedy of the Commons allegory and its corresponding game-theoretic models are capturing anything of the situation here, they are capturing not the moral dilemmas between rational self-interest and the common good, but the limits to our knowledge. Our decisions to consume joint public goods imply no necessary disparity between what we would choose to do in self-interest and what the demands of an impersonal morality would impose upon us.

One might raise the following objection: "Selfishness is not necessarily ruled out by the fact that we cannot practically know when our actions might undermine the well-being of all, including ourselves. Moral concerns would not simply be cast aside even if we had remarkably perceptive human beings who could in fact discern their status as i_k. Consider the person who happened to know where the 'tipping point' (k) was and rushed to consume (or pollute, or whatever) before that point was reached. This individual would seem to be morally challenged and not just ignorant of the consequences of her actions (indeed, as posited, she might not seem ignorant of such consequences at all). And this would seem to be true regardless of her actual proximity to i_k as well as her knowledge of that proximity."

This objection is an important one, and is no doubt rendered all the more plausible by the fact we routinely witness individuals and institutions acting with seeming abandon in depleting or despoiling the natural world in part because it may seem obvious enough to them at that moment that the planet will not collapse on itself as a result of their actions. Notice, however, that the enlightened individual which the objection posits, even if it were possible for her to be certain of the immediate effects of her actions in consuming a public, environmental good, is never actually in a position to make a morally impaired decision. If this omniscient individual is accurate in her assessment of the consequences of her actions, she knows that she will either fall somewhere among the *k*-2 group (i_1 to i_{k-2})—hence consuming the good prior to the point where it becomes a moral matter (since her particular act of consumption will deprive no one of the good)—or she will be one of the participants in the joint public goods game—in which case she is either i_{k-1} or i_k (i_{k+1} to i_N, of course, never get the opportunity to consume the good). Even in the latter case, there are limits to her knowledge, since that knowledge, however complete, can only extend to the current status of the good and the actions that others have already taken, not to anticipating with certainty what actions others *will be* taking. Hence the best that any one individual can do with full information about the "tipping point" would be to distinguish whether she will fall in the *k*-2 group and thus whether her act of consumption has even the potential to threaten the availability of the good at all, or whether she will find herself involved in a joint public goods game with another individual whose joint consumption with her (i.e., mutual defection) would render the good crowded, depleted, or generally unavailable to anyone for further consumption. Since "anyone" in this scenario would include herself, it still seems that this is not a conflict between self-interest and the demands of morality, but rather an epistemic problem of anticipating the actions of $i_{k-?}$ and of coordinating her actions appropriately. Full knowledge, in short, wouldn't be "full" enough, at least not full enough to bring this into the realm of a moral dilemma of the sort that the Tragedy of the Commons proposes.

Even if this were all true, however, and even if this reinforces the notion that what's really at issue in the Tragedy of the Commons and the consumption of joint public goods in general is a coordination problem, this wouldn't rule out the claim that there is some moral issue at stake here. Our skeptic here could still say this: "Sincc this omniscient individual who might know that she is not in the *k*-2 group and hence is involved in the joint public goods game still cannot know what $i_{k-?}$ will do, she still faces a moral dilemma of whether to 'selfishly' consume the good or 'cooperate' and allow for the possibility that $i_{k-?}$ will 'defect', thereby leaving enough of the good such that it may be sufficiently replenished for all to consume once again at a later time (provided one of the critical participants in the next iteration of the joint public goods game makes the same wise decision). You have claimed that what is gained by that last act of mutual defection in the joint public goods game is insufficient compensation to pull the corresponding payoffs (in the lower right-had cell) from the depths of 'least preferred outcome for each' (you argued, e.g., that experiencing the catastrophes of unabated global warning *and* getting few more miles driven rather than walking is not marginally better than environmental catastrophe *simpliciter* in any practically meaningful way). Still, it seems plausible that bald selfishness could drive someone to 'defect' in spite of himself, and that even though he might be undermining his own access to the good in the future, the decision he makes at that point has all of the psychological earmarks of a morally deficient one, insofar as he knows he's getting immediate gratification at the risk of endangering the availability of the good to others—and, yes, perhaps himself as well." This makes sense, and suggests that, even if there is both an epistemic and a moral component to the Tragedy of the Commons (and joint public goods games in general), my general case here would need a nonarbitrary reason for positing the epistemic problem as more fundamental. Is there such a nonarbitrary reason?

Suppose we consider the following claim. The epistemic problem in the joint public goods game is more fundamental than the moral problem in that the two are asymmetrical in an important way: the moral problem requires the existence of the epistemic problem in order to make sense, but it is not the case that the epistemic problem requires the existence of the moral problem in order to make sense. Whether or not the agent is being "selfish," in other words, depends on what he knows about the situation, but what he knows about the situation won't depend on whether he is "selfish"; that is, his epistemic limits won't rest on the moral inclinations he brings with him to the situation. I think this general claim is defensible, and would tend to support my assertion that the more fundamental problem in the Tragedy of the Commons is not one of reigning in selfishness but rather one of coordinating knowledge among rational individuals.

Consider the first scenario, in which the agent's knowledge is the variable in question. If, on the one hand, this agent somehow knows with certainty that $i_{k-?}$ will cooperate and not consume the good, we can expect him to defect.

This would be rational, and it shouldn't really even be considered "selfish," since the agent has rendered himself i_{k-1}, and the good remains available to others so long as no one selfishly makes himself i_k before the good can be restored, replenished, or otherwise re-provided in sufficient amounts. Yet even if he somehow knows that $i_{k-?}$ will defect and consume the good, as our skeptic suggests, he may similarly defect and consume the good, if only in an attempt to get his last shot at consumption. This, I concede, if it were possible, we might well consider "selfish," in spite of the fact that the agent is depriving himself no less than others of future access to the good. Notice, however, the difference here: the same action—consumption of the good—is considered selfish only in the latter instance when the agent knew he was rendering himself i_k and depleting the good. In the case where he knows $i_{k-?}$ will cooperate, his actions are morally neutral at best. Hence whether we count his actions as selfish, and whether we can consider him morally impaired is going to depend on what he knows about the situation.

Consider also the opposite scenario, in which the agent's moral inclinations constitute the variable in question. Suppose, on the one hand, our agent is motivated exclusively by selfish inclinations, and has every intention of consuming the good, come hell or high water. If $i_{k-?}$ does happen to cooperate, the agent's selfish intentions will turn out to be somewhat benign, and moral scrutiny will be shifted to the remaining individuals who might become i_k. If, on the other hand, $i_{k-?}$ defects, we might say of the agent that his selfish motivations did him (and the rest of us) in. What's important here, however, is the fact that, even if the agent had no selfish inclinations whatsoever, and approached the joint public goods game with the intention of acting in such a way that would maximize the long-term benefits of all, he will be no less ignorant of $i_{k-?}$'s intentions than he would be had he been selfishly inclined. What he knows about how $i_{k-?}$ will act, in other words, in no way depends on his moral motivations; he can realistically be expected to remain in the dark about this regardless of whether he is a selfish oaf.

Thus I think it remains plausible to argue that, whatever the moral content of the Tragedy of the Commons and of joint public good games, the epistemic content still prevails as the more fundamental problem. Even if the agents involved act in such a way as to meet our intuitive criteria for what constitutes "selfish," the importance of their actions will rest on what $i_{k-?}$ ultimately does, and if they temper their actions at all, it will likely be in accordance with this knowledge rather than any selfish inclinations they may have.

One final concern bears mentioning. I have indicated in the remarks that opened Part I that we are ultimately looking the question of whether and when it is permissible to require individuals to contribute to the provision of things from which they can be said to benefit—even if that benefit is merely inadvertent and via institutions maintained by others (such as environmental benefits). We have seen that this question is essentially one of the provision of public goods, whether they come about via our contributions or by our withholding our consumption of the good. In these first two chapters, I have argued that, while jointness in consumption may well be a useful criterion in the delineation of public goods, it is not a criterion which is in itself morally relevant to the questions at hand. Our decisions to consume such goods imply no necessary disparity between what we would choose to do in self-interest and what the demands of morality would impose upon us.

One might still raise the following objection: "This entire discussion assumed away any normative concerns over *how much* of the joint public good an individual might *intend* to consume, and came to the circular conclusion that there is nothing of great moral interest in jointness. But it is precisely this that is at issue in the Tragedy of the Commons and in the prisoners' dilemma as well as in environmental issues in which greed compels individuals to take more than their fair share of a good which would be in sufficient supply had no one been greedy, or to consume at all a good towards the provision of which they made no contribution. Thus the individual who consumes more of the good than fairness would allow is acting wrongly *regardless* of his proximity to i_k; it is *this* characteristic that we are talking about when we talk about fairness in the provision and preservation of public goods."

This might be a persuasive and well-intentioned objection. The problem is that it conflates the good itself with the means and institutions by which it is provided—in this case, an implied cooperative venture in which individuals agree tacitly to restrain their consumption of the good to a point where there is enough of it for everyone to go around. This, however, departs from any jointness inherent in the good. If it is indeed possible for the "selfish" individual to take a larger share than everyone else would be able to take and still enjoy the good, the problem is not in the nature of the good, but in the fact that there is nothing to *exclude* the selfish individual from doing this. I will suggest in the remaining chapters of Part I that *this* is what establishes the conditions for our moral concern about public goods: they are goods which cannot readily be provided in a market in which demand and consumption of the good are proportional to one's preferences toward it. Hence the cost of excluding them from anyone who wishes to consume them is prohibitively high, and schemes in which cooperation (i.e., payment) is coerced must be organized in order to ensure the provision of such goods. This is not, however, as we will see, a function of the good's *jointness*.

NOTES

1. The name "prisoners' dilemma" derives from a hypothetical confrontation of two prisoners approximating these conditions. The two are thought to be accomplices

on a crime and taken into police custody. Each is given the option to confess his own guilt in return for a lesser sentence, and neither is told of what the other has done provided that option. "Cooperation" in the titular scenario consists in both prisoners refraining from confession and thus making conviction more difficult. "Mutual defection" is marked by both prisoners confessing their crimes, clinching conviction but amounting to a lighter sentence for each than he would invite had he held out (i.e., "cooperated" with his accomplice) while the other defected (confessed).

2. Ostrom's (1990) "self-governance" alternative might present itself once again at this point: why, we might ask, don't participants in the prisoners' dilemma simply communicate their intention to cooperate so the other player can be reassured he can do the same? The recommendation sounds seductively simple, but we must recall that neither player in a prisoners' dilemma would have reason to believe his opponent would interpret his offer as sincere, and would be better off acting as if the opponent indeed didn't. Communications of this nature, if they are to convey information, would require the very cooperation that the prisoners' dilemma is intended to portray as irrational under the very particular conditions of the dilemma; the assumption of no communication among players is less a formal requirement than a substantive assertion that the ability to express one's "intentions" would not change the outcome. The rational participant in a true prisoners' dilemma will graciously accept the offer to cooperate from his opponent, report that he too intends to cooperate, and then proceed to defect, fully expecting his opponent to do the same.

3. For the sake of simplicity and consistency in comparison with the prisoners' dilemma situation, I have retained the labels "cooperate" and "defect" for the options available to you and to your "opponent" here. Bear in mind, however, that these terms can mean any number of things in respect to the consumption of joint public goods, and what they would mean in a particular case would depend on the nature of the good in question. A "cooperative" action, for example, might mean to refrain entirely from consuming a unit of the good at that point if it were, say, a renewable resource threatened with actual depletion—whereas to "defect" in this instance would be to consume the fateful unit of the good. Another situation might suggest that "cooperation" would mean consumption of a modest unit, while "defection" would suggest consumption of a more substantial unit (though, given what I've argued above, the conditions under which we might have sufficient knowledge to draw such a distinction are likely to be rare). "Cooperation" in the interest of joint public goods need not even require that we refrain from *consuming* something, since refraining from *adding* another unit of pollution to the atmosphere is probably something we would also want to number among these actions. Indeed, we need not even restrict "cooperative" actions to those in which *restraint* is exercised; *contributing* to the maintenance of a joint public good—be it through a monetary donation to an advocacy organization, the volunteering of time, etc.—is likely to be something we would want to capture with the term "cooperation."

4. One might further object that "getting mine now" is an adequate statement of self-interest—that is, that my future states may not even have a preference for the joint public good in question, and that I therefore have no reason to behave now in such a way as to ration it. This would ultimately be indefensible; any reasonable defense of self-interest cannot be time-relative, since, if one has reason to promote one's future aims, one has reason to do so now. On this point, see Parfit (1984:92–95).

5. There are more commonly used labels for this sort of game, though each are, like "joint public goods game," tied to particular examples whose relevance to the strategy they are meant to capture are not self-evident. Hampton (1987) terms them "battle of the sexes" strategies after a somewhat dated example in an early text on game theory in which a husband and wife each enjoy viewing different activities (boxing and ballet, respectively), yet each prefers enjoying the favored activity of the other spouse with him or her to enjoying his or her favorite alone. An even more common label for such game structures is "chicken," derived from the actual game popular among teens in the 1950s in which two cars are driven head on at each other at high speeds until one driver (the "chicken") swerves (leaving the other the victor), both drivers swerve (leaving both players "chickens"), or neither player swerves (leaving both players somewhat worse off than any of the other outcomes). See Hargreaves Heap and Varoufakis (1995:35–37) for a comparison of prisoners' dilemma and "chicken" games.

Chapter 3

The Non-Excludability Problem

In late August of 1970, six hundred thousand people from across Europe and various other locations descended upon an island off of the southern coast of England for a six-day music festival. The third annual *Isle of Wight Festival* was held with the blessings of the local citizens of Afton Down and under the organization of three concert promoters, known collectively as "Fiery Productions." In order to collect a £3 admission fee intended to cover the costs of the production, pay the musicians (some reportedly demanding as much as £250,000), and, presumably, return a profit, the promoters erected turnstiles at a few of the festival grounds' access points, lining the remaining vulnerable positions with two walls of corrugated steel, the area between which was patrolled by police and a private security firm with guard dogs. Undaunted by the measures taken to limit access to a paying audience, several hundred thousand young people camped on the hillside outside the perimeter of the official concert—within earshot of the festival, but far enough removed to be separated from the event. A more vocal contingent of French and Algerian youths among this crowd (dubbing themselves "Desolation Row") challenged the admission fee as the commercialization of the "people's music" and demanded that the barriers be dismantled so that they too could enjoy the "people's festival." The promoters pleaded, in return, that they had spent the last year laboring to bring the festival about, that they had bills and greedy musicians to pay, and that they were merely asking for the minimal contribution from each attendee to help defray these costs. No less than the accounting ledger was laid bare from the concert stage: "We need about 150,000 of you to pay in order to break even," promoter and emcee Rikii Farr announced to the throngs outside the wall. His pleas soon turned into hostility, and went unheeded nonetheless; by the third day of the festival, the walls were breached at several points, and hundreds of thousands poured into the festival grounds. In all, about one-tenth of the over half million in attendance actually paid the £3 admission fee. The promoters went into debt, and many young festival patrons went home disillusioned about the prospects for the "Woodstock generation." As history would have it, this indeed would be the last event of its kind.[1]

Did the inhabitants of Desolation Row have an argument—short of the stated "this is the people's *x*," where x is something which one values and which entails costs to other people, but for which one refuses to pay? "Why exclude," the clever economists among them might have asked, "the lot of us when you yourself say you only need money from a subset to cover your costs? The concert is going on whether we're inside the gates or not. We're not taking anything that isn't already made available. We're not even *crowding* those who have paid, since there seems to be plenty of room for us to enjoy it without diminishing the enjoyment of those already inside the festival walls. Indeed, all of your costs were entailed by the first person to walk through the turnstiles; the production costs of this festival were aimed entirely at making it available to *this individual.* If the festival happens, then the second person through the turnstiles and the last person over the wall have added no additional costs that weren't already anticipated (the hundreds of portable toilets required by health inspectors notwithstanding; we can dig our own holes). Why should we be made to pay if the marginal cost to you of our presence here is virtually nil?"

The promoters, of course, might have had a few words to say on behalf of the £3 admission fee. "While it's true that the additional costs entailed by any one of your lot hopping over my wall and enjoying the show are negligible, we planned this event based on a scale we thought could be paid for by a modest contribution from everyone who would come to enjoy it. We anticipated a crowd of about 250,000, and we booked musicians and prepared the facilities accordingly, under the assumption that each of those 250,000 would shoulder some of the burden of making this event happen. If we were to plan the festival based upon what we could capture from the first person to walk through the turnstiles (since, as you believe, this is the only patron that counts in terms of marginal costs), the third annual *Isle of Wight Festival* would be an accordion player in someone's living room. Don't delude yourself here. The fact that no individual among you has a measurable influence on the cost of putting this festival together does not entail that it can be, in any meaningful sense, a 'free, peoples' festival'. If you want Jimi Hendrix, The Doors, and The Who to come to one spot and perform for you, and if you want the labor and equipment necessary to see and hear them, then you had better be prepared to pay the costs of this. And it seems abundantly clear that one fair and efficient way to distribute these costs is to have everyone pay a small fee to us, the entrepreneurs here who took the chance that 250,000 of you would show up. And, yes, this is true *despite the fact that no single one of you adds anything significant to the total cost of this festival.*"

There's more underlying this dispute, of course, than the stereotypical clash of "establishment" and "counterculture" values which many partisans had invoked to characterize the 1960s. It is as much a question over the provision of a public good as is the question of whether one should wash one's car in times of drought or limit one's greenhouse gas emissions. Unlike the latter questions, however, the question of whether the *Isle of Wight Festival* should have been a free-festival concerns not just the number of individuals that can enjoy the good without jeopardizing its availability to everyone, but *whether* and *how* the good is to be provided in the first place. In this and the next few chapters, I will suggest that such questions more directly provide the content to our moral concerns about public goods.

The business of popular music has changed significantly since the early 1970s, and the technologies of easy digital recording, distribution, and sampling of music have opened up a host of new "public goods" problems that were unforeseen in the era of Jimi Hendrix and that animate legal disputes today between recording artists, major media corporations, and young, computer-savvy listeners. For the benefit of those of us from a generation that can scarcely distinguish an MP3 from an M&M, let's put the question at issue here in the form of a more mundane example: Why is it that concerts, performances, lectures, etc. in auditoriums typically (though not always) require an entrance fee and that parades almost never do? Parades are no less costly to produce; there are often more people and equipment involved in conducting a half-mile Independence Day parade in a small town than in a community-theater production of *Annie Get Your Gun*, yet no charge will be assessed for the parade while at least a nominal ticket price will likely be asked for the musical. Is it for purely symbolic reasons: parades invoke patriotic themes of war and remembrance that would be sullied by demanding a price of admission, while auditoriums house a fare of somewhat more commercial significance? Despite a superficial appeal, his doesn't seem to be a factor; parades can be inspired by crassly commercial themes, while staged productions and concerts can invoke themes of gravity and national importance (recall the flurry of free memorial concerts performed around the U.S. in the wake of the 9/11 terrorist attacks).

A moment's reflection will reveal that parades are merely one step further along on the continuum that distinguished the openness of a park from the confines of an amphitheater for the promoters of the Isle of Wight Festival, and that monetizing and collecting an "admission" fee for a parade forces similar types of trade-offs. There are, indeed, two prominent, conflicting considerations on the continuum from the enclosed theater to the open park to the parade route: each venue exposes a given "production" to an increasingly larger number of people, yet each venue makes it increasingly more difficult to collect an "admission" fee from those same individuals. The concert arena may hold only a few thousand, yet each individual among those would encounter formidable walls and an entrance gate where an agent is waiting to collect her admission fee (assuming an audience that is more or less compliant with walls and turnstiles). The public park might accommodate several thousand more onlookers, but it doesn't come with entrance gates, and access is readily available from all sides in its given state. Any attempt to collect an admission fee would entail further costs in erecting a means to admit those willing to pay the fee and exclude those who are not, as Fiery Productions learned in 1970 (put aside the costs of securing the right to limit a municipal park to private access for the duration of the production). These costs would be exacerbated in a parade setting, if they were retrievable at all. Even if it were feasible to charge admission to city sidewalks and curbs, and to somehow erect a means in which payment could be enforced and nonpayers could be walled out, how would one collect an "admission" fee from those already situated along the parade route by virtue of living or working on the other side of the windows that line the street?

We have already seen that jointness in consumption is characteristic of some goods and not of others, but that jointness alone doesn't seem to entail any normative concerns regarding when it is permissible to coerce cooperation among individuals when the provision of certain goods or the depletion of others are at stake. Note that the jointness criterion is not very helpful in the decision confronted in the current example. Each of the goods described here—an indoor production, and outdoor production, and a parade—carry a certain degree of jointness in consumption, and each may be further described as *crowdable* joint goods (there can literally be "too many" people lining a parade route for you to enjoy it). Yet parades and open-air festivals typically strike us as "publicly" provided goods, while the indoor concert at the other extreme seems "private" enough. What underlies this intuition?

In addition to their strong (though not necessary) tendency toward jointness in consumption, we might further describe public goods as those for which two conditions obtain:

1. *There exists a potential disparity between the population which stands to have access to the good (N) and the minimum number of individuals required to bear the costs of providing it (k).*[2] This is not entirely different from what economists typically describe as characteristic of *positive externalities* (the counterpart to the negative externalities discussed earlier)—the potential benefits which accrue in the production of a good to individuals who do not bear the costs of its production.[3] Note that N may

range anywhere from two persons to the population of the world, while k can vary from a single individual to the world's population minus one (such as in the odd but theoretically possible case where one recalcitrant individual in the hills of Tajikistan benefits from the rest of the world's efforts toward limiting carbon dioxide emissions). Disparities at either margin are certainly less prevalent than the more mundane case of a publicly provided good—where some group k cooperates and bears the costs of its provision, while a somewhat larger group N enjoys its benefits. The important point is that any of these situations differ from the provision of purely private goods, where only those who pay to purchase the good enjoy its benefit.

2. *There exists a cost-ineffectiveness on the part of k to exclude any members of N-k.*[4] This suggests that the problem of excluding non-contributors does not inhere in the good itself but rather is a function of the means by which it may or may not be provided. It is not inconceivable that a secured, mile-long tent might enclose a city street which will host a parade; nor is it impossible to erect such a structure. The point is that it is rarely cost-effective to do so and to then sell tickets to the parade for the purpose of private profit, since the price which patrons would be willing to pay to see the parade *and* to meet the costs of the elaborate exclusion scheme would not likely be enough, in the aggregate, to rationalize the erection of the tent structure in the first place. Similarly, water filtration facilities can construct elaborate valve systems to direct water only to those who have contributed to the production of the facility, yet this likely renders the facility too expensive to build in the first place. Thus the problem of *price exclusion* is less one of pure logistics than of economics.

Taken in tandem, the production of positive externalities and the limits of using price to control access to the good constitute the *non-excludability* criterion of public goods—the counterpart to jointness in consumption. Unlike jointness, non-excludability speaks to not the good itself (and what happens to it during its consumption) but rather to how it is *provided*; it refers not to the benefit but to, as Cullity (1995) terms it, the *benefit-conferring scheme*.[5] Thus what makes public goods "public" in this sense is the fact that, unlike private goods, there is no incentive for the necessary individuals i_1 to i_k to bear the costs of producing such goods; if produced, the good would benefit N—and thus N-k no less than k—and it would be unfeasible and inordinately costly for i_1 to i_k to exclude i_{k+1} to i_N from enjoyment of the good. Hence non-excludable public goods will tend toward sub- or supraoptimal allocation: they will be *under*produced to the extent that, if not provided, some individuals who desire them and who might otherwise be willing to pay for their provision on a market will go unsatisfied; they will be *over*produced to the extent that, if provided, some individuals who would otherwise not be willing to pay for them on a market will still enjoy them (and, if the good is crowdable, deplete them). Since no individual faces a return on such a good proportional to her effort in producing it (i.e., no individual has an incentive to be a member of the set i_1 to i_k), no individual or group of k individuals is prone or able to produce the good on a market, "price" it, and allocate it as if for private consumption.[6]

None of what has been said thus far should imply that the non-excludability criterion provides a neat delineation between goods deemed "private" and capable of being allocated solely by markets and those considered "public" and thus a common source of market failure. Note, for one thing, that just about all goods produce some amount of positive externalities, however small and immeasurable those externalities may be. Many things that we purchase to satisfy a personal desire—be it a musical recording by a preferred artist or our favorite candy bar—makes us a more agreeable person in dealing with other individuals, and those individuals would probably place a positive value on our geniality. Even some things we purchase to treat minor ailments and thus maintain our own health—a box of adhesive bandages or some cold medication—makes us a more productive member of society. We could, in principle, charge our fellow citizens for these things, and demand that the pharmacist or clerk at the convenience store bill society at large for our purchases. But for positive externalities of this sort, too numerous and too evasive to be accounted for in any meaningful sense and likely to be compensated for in other consumers' private purchases of the same items, it doesn't seem worth the effort or cost of maintaining a ledger. We allocate the goods on a private market (in capitalist societies, at least) and let the externalities accrue, come what may. That others benefit somewhat from our purchases doesn't strike us as a failure of those markets; it is still worth the effort for Johnson & Johnson to produce Band-Aids, because we are still willing to pay for them in spite of the fact that others will benefit in some remote way from the healing of our abrasions.

At some indefinite point along this continuum of accumulated positive externalities, the benefits which accrue to individuals who do not bear the costs of the good's production become virtually indistinguishable from those enjoyed by contributors to the good. The additional pedestrian safety secured by a stop sign at a busy intersection is the same safety enjoyed by everyone, both the community activists who labored to see it erected, and the rest of the community who may have been indifferent all along. Similarly, one enjoys the same parade no matter where one stands or whether one contributed to the costs of arranging it. These are not residual, shadow benefits to a good whose primary benefits are bestowed upon purchasers of the good; they are the *same* benefits. For this reason, we consider these goods less prone to market allocation and deem them *public* goods which, if they are to be provided efficiently, must be provided by other means. Yet these goods at either end of the spectrum merely demonstrate that some distinction exists (and that we would be promoting a slippery-slope fallacy to deny this); they do not, in themselves, indicate exactly where and how the line between public and private goods is to be drawn.

Similarly, note that all goods produced for a market entail an irreducible amount of exclusion costs—even this characteristic is not unique to what we might consider to be non-excludable public goods. Ostensibly private retail establishments such as stores and restaurants, De Jasay (1989:128) reminds us, are replete with various exclusionary devices: doors, counters, cashiers, etc. McDonald's could sell hamburgers for a good deal less than it does now if it were to situate new restaurants in open fields, lay the merchandise out on warming trays, and ask that patrons drop a contribution for their meal in the cash box on their way "out." Chances are, however, that McDonald's would not be in business for long under such conditions. Yet neither would it be in business if, after factoring in the costs of erecting walls and counters to prevent us from snatching up as many burgers as we can carry, they couldn't produce burgers at a price we would be willing to pay. Thus it is probably less precise to say that hamburgers are intrinsically "private" and parades inherently "public" than it is to say that the costs of producing hamburgers and preventing just anyone from consuming them can be kept below a price people are willing to pay for hamburgers, while the cost of producing a parade and excluding those not willing to pay to see it are likely to be higher than anyone is willing to pay to see a parade, and thus prohibitive if parades were, like fast-food hamburgers, left to market provision.

Consider another point that may help to clarify the admittedly imprecise distinction between public and private goods, and which was essentially at issue between the *Isle of Wight Festival* promoters and the festival's gate-crashers. In order to sell goods on a market, the producer/seller must be able to assess her costs. If the price consumers are willing to pay for a unit of the good is less than the costs required for the seller to produce a unit of the good, it would be irrational for her to produce it and we would expect no transaction. If, however, the price consumers would be willing to pay to purchase a unit is higher than the cost to produce that unit, the seller can realize a profit and would thus have a rational incentive to produce the good. But how does she determine the cost per unit of producing the good? Take, for example, the tangible case of athletic shoes. On the one hand, the manufacturer may base the price, in part at least, on the *average* cost of each pair of shoes—that is, the total cost of materials, labor, celebrity athlete endorsements, the opportunity costs of foregone production with the same resources, etc., divided by the number of pairs produced. It may, on the other hand, elect to base the price of a pair of shoes on its *marginal* costs—the cost of producing the last unit of the good. For athletic shoes, as for most goods we might reasonably consider "private," these two measures of cost will be fairly congruous; after an initial and unavoidable start-up (and barring depreciation on machinery) the cost of producing an additional unit of a good will not be substantially more or less than the cost of producing the last unit, and both should approximate the average costs of production. Thus the manufacturer's selection of cost on which to base price here is largely arbitrary (though, in practice, circumstances are likely to make one cost assessment more attractive than the other).

Things are different with the concert held in an open park. One could just as readily determine an average cost of the concert per user by dividing the total costs of the performance's production by the total number of people witnessing it; recall that this was Rikii Farr's preferred approach. If the costs of actually collecting an "admission" price (as well as excluding nonpayers) were lower than the gains to be made by doing so, the concert promoters can be expected to hold the concert and charge such a fee. Put another way, if the costs of actually enforcing the fee (collecting and excluding) were incorporated into the fee itself, and if demand remained sufficient for the promoters to capture a profit, we would expect the "show to go on." As we have seen, this is not always easy to assess before the big event. More important, however, is the nature of the marginal costs in this case. Whereas producing another pair of athletic shoes costs about the same as it did to produce the last unit, providing a concert performance to another person to the audience of the concert adds virtually no additional costs once the concert has been produced for the initial audience of size *n*. And for all intents and purposes, *n* in the case of the concert is equal to 1, since all of the costs of the performance's production are the same for the first person as they are for person number 100 and person number 1000, and so on (though initial costs will surely be different, depending on whether the promoters can reasonably anticipate closer to 100 attendees or 1000). Thus after the initial provision of the good, the marginal costs of its provision decrease precipitously to zero, as our imaginary spokesperson from Desolation Row pointed out.

This focus on the decreasing marginal costs suggests that the real problem for the open-park concert producers is not simply excluding nonpaying spectators but rather establishing an admission price to be paid in the first place. The concert producers find themselves in a dilemma: if, on the one hand, the admission price is based on average costs, each individual spectator takes on the risk of being a part of a small audience, hence paying a potentially large "share" of the costs of production—and few individuals could be expected to take on such a risk; hence average cost pricing is likely to exclude individuals otherwise willing to pay a marginal-cost price for the concert (since this cost is not tied to the total number of paying spectators). Yet if marginal costs are to determine price, few individuals could be expected to purchase the first, vastly more expensive, ticket. Even if, as one might suggest, we attempt to "smooth out" the precipitous decline in price by "forcing" a graduated decrease between ticket number 1 and ticket *n*, a schedule of "multiple pricing" is hardly something that can be enforced in the context of a decentralized market (which is, recall, precisely the question here). Given these potential discrepancies between average cost per unit and that same unit's marginal costs, and thus given what J. G. Head (1962) refers to as the problem of *rapidly decreasing marginal costs*, we are able to say of certain goods that their non-excludability ensures that they are difficult if not impossible to efficiently provide through normal market means. We thus have further grounds for delineating *public* goods stemming from *non-excludability* in the ways we provide them.

This ostensibly unavoidable fact—that exclusion costs are likely to discourage private, market provision of public goods—leads some individuals to call for non-market cooperative schemes to provide such goods and an extra-economic, moral authority to coerce contributions from potential members of N-k. This is, I believe, the more fundamental moral content of the public goods problem, and is the source of many of the controversies described earlier which prompted these expository digressions into the nature of public goods. With this understanding of the normative import of public goods, we can now return to the question of our moral obligations in contributing to the provision of "environmental goods" as public goods.

NOTES

1. Murray Lerner's 1995 BBC documentary, *Message to Love: The Isle of Wight Festival*, chronicles the events of these four days in 1970. Editing and post production on the film were delayed for over two decades while Lerner sought backers who would allow the finished product to reflect as much of the conflict and estrangement of the counterculture as the music of the era. Yet even the commercial irony in the music itself is not lost on those of us whose adolescence coincided with the immediate post-Woodstock generation in the early- to mid-1970s. The festival and film feature several lackluster performances by icons of the late-'60s free-festival circuit, upstaged handily by bands such as Free, Jethro Tull, and Emerson, Lake and Palmer, each of whom would come to herald the migration of rock music from the outdoor (and occasionally fenced) festival to the confines of the sports arena—that is "stadium rock."

2. That the notation here is similar to that which was used to formalize the Tragedy of the Commons scenario is no accident. It should be clear at this point that cooperation in the provision of public goods such as environmental quality can come in the form of action—such as contributing money or effort toward a collective cause—or restraint from action—such as refraining from polluting. In either case, some typically

substantial set of persons (N) stand to be affected by the outcome and are thought to have the option to act or refrain from acting appropriately, and some subset (k) will constitute the critical threshold determining whether an existing good is depleted by overconsumption or provided in the first place by sufficient contribution.

3. De Jasay (1989:127–134) aptly describes this as a weak nexus between paying for the good and enjoying its benefits. He, like other economists less prone to describe positive or negative externalities as a market *failure*, considers such situations more as *uncaptured* markets.

4. Until this point, I have loosely adopted the common label for members of the N-k group: *free riders* (this was true even in the previous chapters when *k* referred to the minimum number of consumers necessary to deplete a joint public good). This actually begs the question as to their particular motivations which are, as we will see later, morally relevant here. Given this concern, I will hereafter refer to this group generically as "non-contributors" unless conditions require otherwise.

5. This being said, it's not clear that the same claim would further justify the idealism in De Jasay's (1989) conclusion that "public goods owe their publicness not at all to the 'logistics' of exclusion" but rather to social conventions (De Jasay, 1989:128–129; "it is the public that decides what shall be a public good" he announces in the Preface, p.8). De Jasay concludes from the fact that exclusion costs for schools and health care facilities are relatively low (they *do* have doors, after all, to keep out the riff-raff) that it must be "social mores" that keep them "publicly" provided rather than privatized and made available only to those willing to pay an entrance fee. This seems to miss the point of their "publicness." Access to education and hospitals are *welfare* goods; they contribute to what we like to think of as a person's well-being. They may, as De Jasay quite rightly points out, be provided privately on a market just as easily as they may be publicly subsidized. The somewhat less tangible goods which obtain from living in a society with healthy, educated people, however, are available to every individual in that society, even if he or she is too poor to make use of privately provided schools or hospitals, and excluding such goods from individuals unwilling or unable to pay for them would indeed be a costly venture. It is because everyone stands to benefit from the provision of such welfare goods in this way that health and education may be considered *public goods*—again, insofar as the costs which would be required to exclude individuals according to their compliance with payment schemes would effectively prohibit production of these benefits in the first place. To attribute the "publicness" of such goods to "contemporary mores" which would presumably consider their denial to any individual morally repugnant begs the question at issue: Are moral regimes which demand payment from non-contributors justifiable, given that those non-contributors cannot be denied the goods provided by the cooperative efforts of others?

6. Non-excludability is sometimes referred to as jointness in *production* (to distinguish it from the jointness in *consumption* considered earlier): to make a good characterized by jointness in production available to one person is to make it available to everyone. Occasionally, the goods which jointness in production describes are deemed "pure step goods" or "lumpy goods"—those which "only exist after a substantial amount has been contributed to their production, and then do not increase

in quantity or quality if any further contributions are made" (Hampton, 1987:248). It may be a good idea, however, to avoid collapsing *jointly produced* and *non-excludable* goods into the same category of goods. Hampton cites bridges and the election of a political candidate as examples of step goods, but even these examples hint that the "lumpiness" of a good is independent of its relative excludability. "Half a bridge," as the saying goes, "is no bridge at all," and if the one dollar contributions of 100 people were to purchase the materials and labor necessary for a complete 100-foot bridge, it would certainly be the cases that further contributions would not purchase a longer bridge (though they could purchase a wider one). This doesn't seem to have much to do with the bridge's availability once produced; indeed, it would seem fairly easy to erect a toll gate at each end and demand a bridge-crossing fare from contributors and non-contributors alike. Similarly, it takes a finite (and currently exorbitant) amount of money to elect a political candidate to office, and contributions over a certain threshold will not aid in the campaign but be stashed away in the candidate's coffers for later use. Yet the current state of the U.S. electoral system should be evidence enough that the "benefits" of such goods can be parceled out to campaign contributors and denied to non-contributors with relative ease and little cost to the candidate and her supporters. Goods can be "lumpy" and remain manifestly excludable; that a good may be characterized by jointness in its production should not distract us from the fact that its relative non-excludability is not inherent in the good but a matter of the scheme which provides it.

Chapter 4

Fairness and Non-Excludability

Let's give the label *arguments from non-excludability* to the class of arguments which suggest something like the following: because contributors to public goods cannot preclude non-contributors from enjoying those goods which their efforts and contributions have secured, coercive means to exact contributions from non-contributors are permissible. The structure of such arguments is fairly straightforward; the fact that contributors cannot feasibly prevent non-contributors from enjoying the good is intended to provide, in part at least, a rationale for schemes which force contributions. Cast in the specific terms of the natural environment, we can say that arguments from non-excludability would suggest that it is justifiable to demand that everyone bear the burden of mitigating climate change and other ecological maladies, either in their personal lifestyles or in the taxes they pay, or both, since no one—not even non-contributors—could feasibly be denied the benefit of these things once provided.

There are a number of important things to notice about an argument of this form, not the least of which is the fact that, as stated, it can't work; the argument extracts a normative claim about the permissibility of coercion from the sole descriptive tautology that non-contributors to non-excludable goods cannot be excluded from the enjoyment of those goods, and the history of philosophy is peppered with disputes about the legitimacy of deriving an "ought" from an "is." What is needed is to make explicit the implied premise—the further normative claim which specifies the general conditions of fairness in cooperative ventures. We have precisely this in what has come to be known as the "principal of fairness" in a number of debates among modern moral philosophers. Nozick's (1974:90–95) statement of the principle is appropriately lucid, given that he was launching what was believed to be, at the time, one of the more damaging critiques of it as it was laid out by John Rawls in

A Theory of Justice (1971:108–114) and earlier by H. L. A. Hart (1955), who himself described it as a "principle of mutual restriction":

> This principle holds that when a number of persons engage in a just, mutually advantageous, cooperative venture according to rules and thus restrain their liberty in ways necessary to yield advantages for all, those who have submitted to these restrictions have a right to similar acquiescence on the part of those who have benefited from their submission. (Nozick, 1974:90)

This normative principle, in conjunction with the descriptive observation that contributors can't feasibly withhold the good in question from non-contributors, would thus be minimally required to make the inference to the normative conclusion concerning the permissibility of coerced contributions.

We should also note the relative strength of the claim being defended in arguments from non-excludability. To say that it is *permissible* for contributors to exact a contribution from non-contributors is to say more than non-contributors are *obliged* to contribute. The question of what we may rightfully do with our property—such as withholding it or contributing it as we see fit—is conceptually distinct from the question of whether others are permitted to make demands of it. As David Schmidtz (1991:136–137) has elegantly described, these are independent moral boundaries.[1] To specify the conditions under which one is *obliged* to contribute her property toward the provision of a public good is to defeat one's *prerogatives.* To specify the conditions under which others may *permissibly demand* contribution to a good is to enhance the *restrictions* others may place on one's property. Prerogative defeaters, however, are not necessarily restriction enhancers; it is possible to argue that one is indeed obliged to contribute and still maintain that others are not permitted to exact that contribution by force. Notice, however, that the opposite is not true. If we may adequately argue that others are *permitted to coerce* a contribution from an individual, by implication, we are arguing that she is *obliged* to contribute as well. In specifying the conditions for permissible coercion, we have effectively defeated the prerogative of the non-contributor to withhold that contribution. While it wouldn't be nonsensical to say, in other words, that though I'm obliged to contribute *x* toward the cooperative scheme that you've designed, you are not permitted to take it from me by force, it *would* seem nonsensical to say you are indeed permitted to take from me something which I'm not obligated to surrender.[2] Thus, when arguments from non-excludability suggest that contributions may be permissibly exacted because the benefits of the good may not be practically denied to non-contributors, they are also suggesting that non-contributors are *obliged* to contribute. Since the opposite case cannot be said to be true (i.e., obliging *x* would not entail permissibly coercing *x*), we can say that arguments from non-excludability as described here are more ambitious than they would be if they were attempting to justify only the obligation to contribute.

Note also that the stated desirability of the non-excludable public good to non-contributors does not matter in arguments from non-excludability. That fact that non-contributors could, in principle, enjoy the good if they availed themselves of it is taken as more or less sufficient reason for exacting a contribution from them; it doesn't matter whether they claim to desire the good or not, or whether they are inclined to seek it out. Thus the sunbather who claims to desire no more than to enjoy a paperback novel in the park is charged no less than those who have convened there for an open-air concert; his professed indifference to the music, sincere as it may be, should not persuade those collecting the admission fee to allow him in free of charge. While this point may seem a mundane corollary to arguments from non-excludability, it has severe implications for our economic understanding of the goods in question: if we say that a contribution—whatever its magnitude and however it might be determined—ought to be exacted for non-excludable public goods *regardless of one's stated preferences for that good*, we tend to preclude those willingness-to-pay measures so dear to economists as the sole determinant of the value of such goods. This makes sense if we consider the distinct nature of public goods described in the previous chapters. Unlike the case of private goods, market participants can behave strategically in securing their access to public goods—and they are likely to be successful in their strategy. Claiming that I am only willing to pay twenty dollars for athletic shoes in a market where the going rate is eighty dollars will not get me a pair of athletics shoes for twenty dollars, even if, in truth, I would be willing to go as high as eighty dollars. On the other hand, were there no means of exacting a "fair" payment for public goods—whatever "fair" might be—all but the rational altruist would be expected to quote a lower offering price for such goods than they actually would be willing to pay were it possible to allocate the good privately. If this assuming-away of subjective, willingness-to-pay assessments strikes one as flippant and premature, it probably should at this point. The free-market libertarian appeal to a purely subjective valuation of goods (as well as the worries over *paternalism* that often heightens that appeal) will emerge again later in Part I, and I will require the bulk of Part II to argue why such measures are not sufficient to assess the value of public, environmental goods. Suffice it to say for now that arguments from non-excludability do not rely on reports from the non-contributor to determine whether he is likely to consume the good in question; they simply proceed from the assumption that he can (which is not to say, as we will see, that one would never have reason for assessing the presumed intentions of the non-contributor).

A further point of clarification: while the duty for any individual (i) to contribute (implied, as we have seen, in the permissibility of coerced contributions) should probably not ignore entirely the number of other potential contributors (else the duty in question might become simply "provide the good" and would be an unreasonable demand if no one else had contributed), it makes sense only in reference to the fairness of the contributions in question, and *not in reference to the number of other contributors*. Put differently, the duty to contribute that arguments from non-excludability propose is not a duty whose moral weight for any individual varies with the degree of compliance by other individuals. Arguments from non-excludability thus abstract somewhat from the actual number of individuals involved (N), the number of actual contributors among this total (m), the number of contributors that would be required to provide the good (k), and any discrepancy between (m) and (k). This is true largely because hinging the moral force of the duty on these contingent numbers is likely to render it self-defeating It would make the duty to contribute a *defeasible* one in that, while everyone would be initially bound by it, the duty could be fulfilled before everyone has made a contribution. Once the public good is secured, in other words, the obligation to contribute to it would be effectively called off, allowing every individual from i_{k+1} through i_N to evade the obligation. Yet if it were possible to escape the weight of the obligation in this way, every individual would have an incentive to be a member of the group N-k, and each could rightfully point out that our demands that each contribute would arbitrarily and unfairly preclude her from impunity in the N-k group.

"No problem," we might think, "We can avoid this dilemma (and salvage the compelling intuition that the degree of compliance by others determines whether or not we are obligated) simply by rendering the relevant duty: '*be a member of a group k* which provides the public good'." This wouldn't do much to bail us out, however, since this duty would itself not take effect until k-1 individuals had, for whatever reasons, already contributed. Individuals i_1 to i_{k-1} would not be, regardless of their intentions, fulfilling the obligation; since the good is not yet provided, they have not made themselves members of a group which acts to provide the good. Individual i_k, as we have seen, is the first who could make a difference in obtaining the good, and thus has the potential to at once transform i_1 through i_{k-1} into members of the morally relevant group. Knowing this, and believing, however accurately, that they are not i_k, everyone could then reason that their contribution would not make them a member of a group (k) which would collectively act to provide the good. We would thus be right where we were before, with i_1 through i_N in the position of being able to shirk the duty with impunity. It might be possible to extrapolate from earlier arguments that it would be ultimately nonsensical for one to hold that her contribution could not be the critical one, and that she is thus not i_k; *someone*, as we saw, has to be the decisive agent whether she is aware of it or not. Still, we also saw that it would be *practically* impossible for i_1 through i_{k-1}—and hence i_k—to be aware of their status even if i_1 to i_{k-1} had made contributions (an action which would seem to be, in this case, supererogatory), thus arguably making the duty "be a member of a group *k* which provides the public good" epistemologically too demanding to be morally binding. We are thus left with the vague but defensible assertion that the contributions under consideration in arguments from non-excludability are determined by criteria of fairness which do not depend on the degree of compliance by others (or which, perhaps more accurately, assume full compliance).[3]

With these points in mind, arguments from non-excludability may be more vividly clarified by way of example. Consider one of the more prominent applications of this form of argument: the rationale for the union shop. Right-to-work advocates often claim that union benefits should be purchased like a commercial good: if one desires the advantages of collective bargaining, one can contribute to its cause; otherwise, one is under no obligation to make such a contribution by joining the local union and paying its dues. It would thus be illegitimate, they argue, for unions to coerce such contributions as a condition of employment by requiring that all employees in a particular workplace (or in a particular industrial sector, if wages and benefits are established industry-wide) join the union. Union advocates respond by pointing out that it really doesn't matter whether the potential employee professes to seek or ignore union benefits secured by collective bargaining: they will have access to them by virtue of taking the job—an access which cannot feasibly be precluded by dues-paying union members. It's not possible, in other words, to erect some form of wage and benefit partition to ensure that non-union employees won't reap the advantages secured by the history of collective bargaining in that industry. Even if it were possible to construct separate pay scales for union employees and non-union employees, it would presumably be difficult to determine on what such scales would be based. Pure market forces—that is, the meeting of schedules of what employers are willing to pay and what non-union employees are willing to accept—wouldn't work, for what you as a rational potential employee would be willing to accept is based on your assessment of what you would need to purchase with your wages, and the prices of what you need to purchase (indeed, what would appear cost-effective in the eyes of entrepreneurs to produce at all) are determined, even if in small part, by prevailing wages in your industry and across industries—many of which already unionized.[4] Hence as long as some wages and benefits are the result of collective bargaining, it would be difficult for the union-eschewing employee to say "What I earn is just my value on a labor market; I am not benefited one iota from collective bargaining in my industry or any industry."[5] *Even if* we could somehow establish a "pure" labor market parallel to our system of collective bargaining, this would address only a portion of the benefits secured by union activity. It becomes a strain on the imagination to visualize how a workplace would arrange to subject non-union employees to health and safety risks that have otherwise been diminished or averted on behalf of unionized employees. Do we, for example, require that non-union chemical workers wear flammable clothing and carry matches to cancel out the fire-safety benefits they enjoy and for which their unionized colleagues had fought? Shall we wax the scaffolding of non-union window washers to cancel out the safety benefits now built into their equipment? It is, in short, difficult if at all possible to abstract oneself from the benefits provided by collective bargaining and other union activities, whatever one might think of them.

In a brief but celebrated passage from *Anarchy, State, and Utopia* (1974:95), Robert Nozick argues that the principle of fairness as stipulated by Hart and Rawls is unacceptable, and that attempts to exact payments from individuals who are provided benefits they do not request are patently indefensible:

> On the face of it, enforcing the principle of fairness is objectionable. You may not decide to give me something, for example, a book, and then grab money from me to pay for it, even if I have nothing better to spend the money on. You have, if anything, even less reason to demand payment if your activity that gives me the book also benefits you; suppose that your best way of getting exercise is by throwing books into people's houses, or that some other activity of yours thrusts books into people's houses as an unavoidable side effect. Nor are things changed if your inability to collect money or payments for the books which unavoidably spill over into others' houses makes it inadvisable or too expensive for you to carry on the activity with this side effect. One cannot, whatever one's purposes, just act so as to give people benefits and then demand (or seize) payment. Nor can a group of persons do this. If you may not charge and collect for benefits you bestow without prior agreement, you certainly may not do so for benefits whose bestowal costs you nothing, and most certainly people need not repay you for costless-to-provide benefits which yet others provided them. So the fact that we partially are "social products" in that we benefit from current patterns and forms created by the multitudinous actions of a long string of long-forgotten people, forms which include institutions, ways of doing things, and language . . . does not create in us a general floating debt which the current society can collect and use as it will. (Nozick, 1974:95)

Nozick's argument here is taken to be one of the definitive attacks on the fairness principle and thus any argument which might rest upon its veracity—as do what I am calling *arguments from non-excludability*.[6] He is arguing by analogy that benefits bestowed upon individuals through the actions of others, unless such benefits are the product of a prior agreement, do not require payment. The general case is built around what he intends to be an intuitively obvious example of one individual inadvertently benefitting another in the pursuit of the former's interest by heaving books through the latter's window; it should be agreeably clear in such cases, he is suggesting, that any demand of payment is morally indefensible. From this example, he concludes that the types of benefits individuals receive as a consequence of group efforts to which, either by choice or by design (as in historical benefits) they are not a party are similarly exempt from the obligation of payment. This would presumably cover a wide array of things typically described as public goods, from union benefits to public libraries, to mass transit, to national defense, and certainly to the sorts of environmental goods that are our principal concern here. It would not be an exaggeration to say that Nozick's argument represents the prototypical libertarian case against taxation for anything other than those goods which people have expressly consented to be purchased collectively on their behalf from their voluntary contributions to a common pool. It is essentially just this now-classic libertarian argument that I am calling into question in Part I, and committed environmentalists may find it unsettling how intuitively appealing a skilled philosopher such as Nozick could make the case sound.

In spite of any superficial appeal in Nozick's argument against the principle of fairness, and in spite of its prominent role in libertarian arguments against cooperative regimes of many sorts, it's worth pointing out that the argument simply doesn't accomplish what it claims to do. While there are undeniable similarities between the individual showered unexpectedly with books through his window and one imparted the benefits provided by past collective actions such as wage and benefit bargaining (or, for that matter, environmental regulation), we cannot readily conclude that what would seem morally objectionable in the case of the former would necessarily be so for the latter. This is not merely because Nozick's analogical case may strike us a preposterous and improbable; this is precisely why philosophers posit arguments by analogy: what seems obvious in the contrived and often bizarre analog can then be generalized to the primary case at hand where real-world details often muddy the situation. Nozick is simply doing his job here. The problem in this case is with *dis*analogies—characteristics of the analog that cannot be said to be true of the primary case at hand, such that we cannot readily generalize conclusions from the former to the latter. Nozick invites us to accompany him on a fairly significant inferential leap when he describes the simple case of the book thrower and his beneficiary and jumps to the case of "current patterns," "multitudinous actions," "long-forgotten people," and "institutions." There are, conceivably, sufficiently important differences between the two types of circumstances which would render Nozick's argument by analogy inapplicable (or, at least, decidedly less applicable) in the case of an individual such as the one indifferent to the gains acquired by her union or by the environmental movement.

Most obviously, Nozick's analogy of the "book thrower" really only applies to private transactions between individuals: one person is quite literally thrusting a "benefit" upon another person in real time and demanding payment for it. To appreciate the difference, note that the book-throwing example would need to be retrofitted extensively to conform to the conditions of the bestowal of the non-excludable benefits that accompany public goods such as union benefits or environmental quality. As presented by Nozick, the analogy of the unsolicited books doesn't line up with cases like union shops, where collective benefits have been secured by the contributions of many people long before I, an indefatigable anti-unionist, come along. The proper analogy here would be not a showering of books into a home I already own, but rather my opportunity to purchase a house which comes with a well-stocked library. The seller may wish not to separate this feature, and certainly has a right not to do so, since the costs she would have to bear in removing the books may still not be covered by my bid for the house. It would seem odd to say that I, in paying for the house and the books, am being forced to pay for a benefit I didn't solicit. I'm perfectly free to reject the selling price and look elsewhere for a home, just as a potential employee is free to reject the conditions of union contributions attached to a job and look elsewhere for employment.

Ironically, both supporters and critics of Nozick's position point routinely to the example in the passages that immediately precede the book-thrower analogy in *Anarchy, State, and Utopia* (pp.93–94) to explicate his position on the provision of particularly *collective* benefits. In this example, the cooperative scheme is a community-run and organized public broadcasting system which requires that individual "participants" (everyone within earshot of the broadcasts) surrender one day per year to take their turn administering the broadcast facilities. While this example more closely approximates the conferring of "benefits" from a *collaborative* scheme than does a solitary individual hurling books through one's window, Nozick clearly intends the latter example to be as applicable to the case of collectively provided goods as the former. Though I've omitted a few sentences in the extended passage above, there is nothing elliptical in Nozick's launch into the final sentence of the passage. Note, moreover, that Nozick's example of the public address system is just that—an *example* of a cooperative scheme in which demands are made upon individuals who cannot be easily excluded from the good provided. His purpose in the book-thrower case, on the other hand, is an *argument by analogy*. It is our assent to the implied conclusion in the analog (that the book thrower would be unjustified in demanding payment for the books)—an assent that should rightly be granted—which is intended to justify the similar conclusion in the primary cases of interest: those where cooperative schemes confer benefits and make demands upon individuals without their consent. If we accept the conclusion in the book-thrower case, he is arguing, we are compelled, by weight of the analogy, to accept it in the case of cooperative endeavors such as the public address system. The example of the public address system itself does not appear intended to carry the same argumentative weight, since Nozick himself makes a separate effort in these passages to justify his conclusion that one is not obligated to take one's turn at the microphone, and it is not a terribly effective one (Simmons, 1979:118, for example, characterizes these passages as relying more on "the snowball effect of deviously arranged examples than on argument"). The point here is that, given the libertarian weight that Nozick's argument from the book-hurler analogy is intended to carry in casting doubt on the legitimacy of coerced participation in collective-benefit schemes (basic government taxation or union dues being prominent examples of these), and given its intuitive, superficial appeal, it is striking how easily it begins to totter under closer scrutiny.

None of what I've said here, of course, provides a knock-down defense of arguments from non-excludability or the fairness principle on which it rests; this is not the primary aim of this chapter or even of Part I in general. It does suggest, however, that given minimal clarification, arguments from non-excludability are at least coherent and have a superficial plausibility to them—certainly a plausibility no less ringing than Nozick's entrenched libertarian arguments. More important, I hope to have clarified one route a skeptic might take in arguing *against* arguments from non-excludability (however much it misses the mark): accepting the non-excludability of the good, but disputing the claim that fairness permits us to demand contributions toward its provision. This will prove important to distinguish other types of objections one might raise against arguments from non-excludability—counterarguments which concern the relative avoidability of the good in question, and the intentions of the non-contributor compelled to "enjoy" it. It is these types of arguments that is our main concern in Part I.

What if, say, a potential employee at a union shop were to argue the following: "Nozick's argument aside, the problem here is not that I can't be *excluded* from the benefits secured by collective bargaining but that I can't be *excused* from them—whether or not I would prefer having these benefits to not having them, I would be compelled to 'enjoy' them if I took the job. Hence I'm not really a 'free rider' on the cooperative schemes of others, I'm a *forced* rider, and this is a morally significant difference. Even if one could justifiably coerce a contribution from the free rider—that is, the individual who truly appreciates the good and can't be precluded from it should he decide to consume it—this does not entail anything similar for those of us who are genuinely indifferent to the good and averse to paying the costs it entails. Indeed, it is because of this in*excus*ability that I should not be compelled to contribute toward the good." Is this the makings of an adequate defense against (at least some) arguments from non-excludability? I will ultimately argue that it is not.

NOTES

1. Similarly, Nozick suggests that an "argument for an enforceable obligation has two stages: the first leads to the existence of the obligation, and the second, to its enforceability" (Nozick, 1974:93).

2. This is an asymmetry that Schmidtz himself fails to acknowledge; in suggesting that the "relocation of one" of these moral boundaries "does not entail the relocation of the other," he appears to imply that it would indeed be plausible to say that you are permitted to take from me what I am not obligated to voluntarily surrender.

3. Note that Liam B. Murphy (1993), in one of the more eloquent recent defenses of the notion that an "agent is required to sacrifice only as much as would optimally be required of her under full compliance"—what he calls the Cooperative Principle—concedes that there are situations where our duties do indeed seem to hinge upon the degree of *actual* compliance by others. I paraphrase loosely one such example he describes: You are driving to the airport to meet a non-refundable flight for a much-needed vacation. You see a small child flailing in a drainage ditch off of the access road you are traveling. Pulling over to save the child, you acknowledge, will mean missing your plane and thus your vacation, but you conclude reluctantly that providence is not on your side, and that you are obligated to save the child. A few months later, you have reserved a similar flight for a similar vacation. Once again, you notice from the access road on which you are driving that a small child is flailing away in the nearby drainage ditch. This time, however, the child is joined by another small child in the same predicament. This time, moreover, you are joined on the road by another motorist following close behind you, also arriving to meet a non-refundable flight. The situation is such that, only if both you and the other motorist pull over and save one child each, you would each have sufficient time to make your flights. In fact, however, the other motorist does not pull over, and drives merrily on his way to meet his plane reservation with time to spare. The Cooperative Principle would here seem to suggest that you could pick one child at random, save her, and leave the other to drown with impunity, since your obligation is only what would be required of you had the other motorist complied with his apparent obligation to save one of the two children. This, as even Murphy acknowledges, seems morally reprehensible. Drowning-child scenarios, bizarre as they often sound to non-philosophers, have an extensive history in modern moral philosophy and are often useful in highlighting specific aspects of moral dilemmas. Peter Singer was among the first to employ them in his germinal essay "Famine, Affluence, and Morality" (1972) to similar effect: our moral duties are not mitigated by the fact that there may be many other people who could fulfill that same duty.

4. Anti-unionists can and do argue here that this is precisely the problem of labor unions: everyone is disadvantaged by them to the extent that they introduce wage

inflation into an otherwise stable labor market, initiating an upward spiral of wages and prices and generalized economic inflation. The issue is probably best left to demonstration of the formal economic models of this by a competent economist, but if there is any intuitive plausibility to the claim as stated, there is probably an equally intuitive plausibility to the counter-argument which would suggest that *everyone is benefited* by the presence of unions and the intervention of collective bargaining into otherwise "pure" labor markets: as long as there are fewer jobs available in a market capitalist than there are individuals willing to work, unions avert an economic crisis of underconsumption by preventing the downward spiraling of wages that would be witnessed in "race to the bottom" of a pay scale and the ensuing inability to pay a market rate for the products they produce, forcing down prices, and generally deflating the economy. In a very general sense, this is the principal concern of Keynesian and other "demand-side" economic theories.

5. Lowe (1996) argues that some individuals are not only *not advantaged* by the collective bargaining efforts of unions, but actually *disadvantaged.* He proposes as an illustration the exceptional philosophy professor who so outshines her colleagues that it would be accurate to say no part of her relatively higher salary can be attributed to the collective bargaining efforts of the university teacher's union to which she is forced to pay dues. Indeed, Lowe fears, a collective bargaining agreement may inadvertently put a ceiling on her pay which might otherwise be offered by a university. While I don't find, from an economic standpoint, Lowe's arguments persuasive (it's not at all clear how union collective bargaining necessarily subverts merit compensation), the more salient problem with Lowe's case seems to be a prudential one. While it is all but inevitable that some laborers in any field—be it lemon-picking or philosophy teaching—will outperform others, and that their performance would fetch higher compensation on a pure labor market, one cannot know ahead of time that one's skills are of this caliber. One rationale for participation in a collective bargaining agreement would thus be to insure against mediocrity from behind a "veil of employment ignorance"; since I can't be certain that my skills are going to make me outshine the rest, it would be in my interest to make certain that "the rest" are compensated at a rate that would be adequate for me. While this is not a moral argument *per se* (despite its Rawlsian spirit), and thus may not address the fairness of coerced contributions to union efforts, it does suggest that it would be irrational for anyone entering into a labor market to excuse oneself from union dues solely on the basis of confidence in one's abilities.

6. See, for example, Bell (1978), Simmons (1979:118–142), Arneson (1982), Cullity (1995), and Lowe (1996). These merely scratch the surface, however, of a much wider body of literature on Nozick's treatment of the fairness principle.

Chapter 5

Non-Excludability and Non-*Excus*ability

Let's return to the example with which we began Part I. Both Smithers and Burns are confronted with an opportunity to consume a good provided by the cooperative efforts of others—a clean and safe water supply resulting from a new water filtration facility. The good might be said to be not a frivolous one by most sensible people, insofar as the water supply prior to the building of the filtration facility was poor, and prone to occasional risks in water-borne bacteria, most of which could be all but eliminated by the filtration plant. The good can also be described as non-excludable; both Smithers and Burns—and anyone else who might choose—can consume the clean water whether they have contributed to the provision and maintenance of the filtration facility or not. Indeed, it might even be said that the good is unavoidable, given that this is the only supply of tap water in town, and one would have to take exorbitant steps to secure another source of water. The only difference between Smithers and Burns appears to be one of preference. Burns likes the idea of cleaner water, relishes its provision, and thus can be fairly said to "benefit" from the efforts of others, free riding on the good. Smithers, on the other hand, is genuinely indifferent to the clean water and consumes it only because it is the only tap water readily available to him.

Should Smithers and Burns, given these general conditions, and given what has been discussed so far in these chapters, be treated differently? A number of possible arguments present themselves concerning their relative moral status:

I. Burns may well be deemed a "free rider" in this example, and, if so, may be permissibly compelled to pay his fair share in the provision of the good. Smithers is clearly not a free rider since he has no desire for the good, and would prefer not receiving it to receiving it and contributing toward its provision. Smithers should thus be exempt from any compulsory payment scheme to which Burns might be deemed accountable. This argument asserts that individuals consuming public good may have different desires and preferences and that these are sufficient in deciding the scope of the obligation to contribute. Note that both the libertarian and the environmentalist might be quite content with this situation; as long as we can discern those true desires and preferences, it seems fair to collect contributions accordingly. The difficulty, of course, is in discerning those desires and preferences, since everyone would have a reason to claim indifference—some sincerely, others not so sincerely. We'll thus call this the *argument from preferences* for our purposes here.

II. Because the good is unavoidable, Smithers should be released from any compulsory payment scheme. Even if arguments from non-excludability may be said to be justified, they must acknowledge this potentially important characteristic in the nature of the good provided, since that good may be such that one *cannot help* but consume it, and it would be wrong to compel people to pay for something they cannot help but consume—whether or not it is something they actually desire. This argument asserts that there is something in the nature of the good that determines whether or not a compulsory contribution scheme is justifiable—namely, the degree to which one is "forced" to "enjoy" it. And if Smithers can be justifiably released from such a scheme by virtue of this argument, so too must Burns enjoy the same privilege, despite the fact that he does indeed prefer having the good to not having it, and would surely "free ride" even if he weren't compelled to enjoy it. We will, for reasons to made clear shortly, call arguments which release from compulsory contribution *all* consumers of goods from which they cannot readily be exempted *arguments from non-excusability*.

III. Burns is clearly a free rider in this example, and should thus be compelled to pay his fair share in the provision of the good. Yet, as the argument from preferences (I) concedes, we have no means of distinguishing Smithers' preferences and desires from those of Burns, and while Smithers may report that he has no interest in the good, he, no less than Burns, has an incentive to misrepresent (or even miscalculate) the value he holds for the good. It is therefore because the filtered water is a non-excludable good that both Smithers and Burns should be considered, like all non-contributors, free riders, and should be compelled to contribute out of fairness. *Non-excludability* thus justifies the overriding imperative to contribute, outweighing any competing concerns about differential preferences and intentions or the relative unavoidability of the public good in question; non-excludability alone is, in short, sufficient to justifiably exact contributions from *all* potential non-contributors. We'll call this the *strong argument from non-excludability* to distinguish it from the argument as I have presented it thus far, which we should now deem the *simple* argument from non-excludability. The simple argument from non-excludability appealed only to the public nature of the good and the impracticality of withholding it from non-contributors. This strong argument from non-excludability appeals further to our epistemic limits in distinguishing morally the ill-intentioned from the well-intentioned non-contributor—that is, those who desire the good but wish to free-ride on others contributions from those who are genuinely indifferent to the good and don't care whether others cooperate to provide it. It is thus concerned with *fairness* in the distribution of the costs of providing the good in a way that simple arguments from non-excludability are not.

These three argument types certainly do not constitute an exhaustive list of all possible moral positions that might plausibly be drawn from the example, but they do present a few of the more prominent ones. In the present chapter, I will argue that in spite of a strong intuitive pull, arguments from non-excusability such as the one outlined in (II) are ultimately quite difficult to defend, and that they tend toward circularity by appealing to the desires and intentions of agents consuming public goods. In the next chapter, I will argue that the further appeal to arguments from preferences such as the one outlined in (I) are themselves inadequate to vindicate "honest holdouts" (to use the label coined by Schmidtz, 1991, and to which we've already alluded[1]) by morally distinguishing them from free riders. There are limits, as we'll see, to what we might conclude from this about the strong argument from non-excludability, though certain important implications would indeed follow if our libertarian intuitions can be shown to be without much of a foundation. A great deal more remains to be said, however, before even these tentative conclusions can be drawn, especially with regard to just what *non-excusability* means. This brief overview should serve as nothing more than a signpost indicating where we are heading.

The account of public goods I've outlined thus far in Part I has been, however tedious, somewhat incomplete; there is yet another distinction we might draw relevant to the goods in question. We have seen that a good can be non-excludable in its provision to the extent that, once it is provided, it cannot feasibly (i.e., cost-effectively) be withheld from any individual wishing to consume the good, regardless of whether that individual contributed toward the good's provision. For a subset of non-excludable public goods, an additional characteristic may be said to describe them: once provided, a definite set of individuals *must* consume them—that is, there is no discrimination in their consumption, again, regardless of whether the particular individuals contributed toward the costs of producing the good. These goods we might describe as non-ex*cus*able.[2] Economists occasionally refer to such goods as *pure* public goods, insofar as there appear to be no means of disaggregating their consumption in accordance with individual preferences. The blanket of national defense provided by a standing army, air force, etc., remains the classic example of a non-excludable, non-*excus*able, pure public good: while one cannot feasibly be excluded from the protection one's national armed forces provide, neither can one escape it, short of emigrating, in which case one is no longer a member of the population for which the good is provided. We can add to this category a number of the environmental goods of primary interest to us here. Clean air, stable global temperatures, a safe and plentiful local water supply—each of these may be described as goods whose benefits both cannot be excluded from any member of the relevant population to which they are provided (arguably, the entire planet, especially in the case of many environmental goods) *and* from which no individual can readily excuse herself.[3]

Just how non-excusable goods might pose a unique moral problem in regard to their provision should be apparent. In the case of non-*exclud*able public goods, we operate on the assumption that free riders enjoy the good, even if they claim indifference and that their access to the good is incidental to their pursuit of other goods; since they have the option to eschew the good, their consumption of it would indicate that they *chose* to consume it, with the knowledge that they were "getting something for nothing." Hence, as we saw, the (simple) argument from non-excludability is advanced without regard to the stated intentions or preferences of the non-contributor, and essentially assumes the good in question is "enjoyed." With goods we might deem non-excusable, however, the situation is somewhat different. If we describe consumption of the good as non-voluntary—even compulsory—we lose the ability to infer that one who consumes the good does so willfully, thus "enjoying" it. Hence whatever moral conclusions we may draw about the relative unfairness of the actions of a free rider would not seem immediately applicable to anyone consuming a non-excusable good but who similarly refuses to contribute toward its provision. While one may desire the good, and prefer having it to not having it, it's also possible for any one consuming it that she is entirely indifferent to it, and would eschew even the possibility of obtaining it via free riding were it possible. These individuals, we might believe, should be morally distinguished from free riders, insofar as we would not want to say of them that they are acting unfairly should they withhold their contributions to whatever cooperative scheme may provide the good. To return to our original example, Smithers, unlike Burns, has genuine indifference for the purified water that is being thrust upon him, which he would rather not receive if it were something for which he had to pay. We might thus conclude from this that it would be impermissible to coerce contributions from individuals such as Smithers, even if it were somehow possible to identify true free riders and permissible to exact contributions from them for non-excludable goods. These individuals—"honest holdouts," as I've been referring to them—do not appear to be "getting something for nothing" in any meaningful sense, nor do they appear to be exploiting the efforts of others for their own gain as does the classic free rider. They are, if anything, "forced riders" on the non-excusable good.

If this is essentially what is being argued in (I) above, note how easily our epistemic limits in distinguishing free riders from honest holdouts can pull the case toward (II), the argument from non-*excus*ability, which would effectively exculpate anyone from the duty to contribute toward the good: "A moral distinction between the free rider and the honest holdout is well-intentioned but superfluous: as long as a good meets the conditions of non-excusability, we should be freed from the obligation of contributing to its provision regardless of our preferences, motivations, desires, intentions, etc. Splitting hairs over the dispositions of agents is unnecessary if those dispositions will play no role in determining whether or not they will consume the good. It is only the nature of the public good itself which determines this in the case of non-excusable goods." If, in short, we accept the argument from preferences (I) and assent that the free rider and the honest holdout are morally different, we might readily be nudged to also accept the argument from non-excusability (II), since the non-excusable nature of many public goods renders the moral difference between the free rider and the honest holdout irrelevant—a difficult question of authentic psychological dispositions that we don't need to resolve.

This might strike those committed to the strong argument from non-excludability (III) as an arbitrary maneuver, however. If we can all agree that the argument from preferences (I) is touching on *something* important in suggesting that the free rider and the honest holdout are of a morally different status, why take the epistemic limits in distinguishing the two as a justification for releasing everyone—honest holdout and free rider alike—from the duty to contribute toward the good just because the good cannot practically be avoided once obtained? Why would it be any less legitimate to conclude the opposite from those epistemic limits: since we cannot practically distinguish free riders on non-excludable public goods from forced riders on non-excusable goods, everyone must be made to contribute regardless of his or her stated preferences. The libertarian, in short, wants to move from (I) to (II) because the free rider is too readily conflated with the forced rider; the environmentalist wants to move from (I) to (III) because the honest holdout cannot effectively be partitioned from the free rider. Both moves seem equally capricious on the surface. Can anything help us to arbitrate a fair distribution of costs and burdens despite our epistemic limits in distinguishing free riders from honest holdouts and from forced riders?

The environmentalist's concern in preferring the move to the strong argument from non-excludability (III) is, as described above, one of fairness and not particularly troubled by the non-excusable nature of the good itself. Non-excusable goods, recall, are also necessarily non-*exclud*able goods, and if everyone were freed from the obligation of contributing to non-excusable goods due to the nature of the good itself, some individuals might still meaningfully be said to be free riding on the efforts of others (if indeed there were enough people cooperating to provide the good in the first place). While these individuals receive no more or less of the good than do those who are genuinely indifferent to the good, they are receiving something which they understand to benefit them yet for which they refuse to pay. Both Smithers and Burns receive cleaner water in our original example, yet it might seem premature to vindicate Burns' actions—which even he might agree are "unfair"—based on the fact that Smithers is compelled to consume the good. The difference between Smithers' actions and Burns' actions seems to lie outside the nature of how the good is provided (though this will surely have something to do with the moral content of our actions). Hence drawing the relevant normative line here at the good itself—that is, non-excludable goods may permit coerced contributions but non-excusable goods may not—will tend to misdirect us from the moral distinction that arguably needs to be made: what renders the honest holdout different from the free rider?

"Still," the persistent libertarian might object, "even if we lose the ability to condemn free riders on the basis of their preference for the good they consume without contribution *whenever the good is also a non-excusable one*, this is a necessary moral concession we must make in the interest of those who remain genuinely indifferent or even averse to the good. A fuzzy moral line which permits some unfair actions in the protection of well-meaning individuals is preferable to one which ultimately risks over-restricting the actions of honest holdouts; it would be better to err on the side of allowing free riders to flourish in the case of non-excusable goods than on the side of forcing individuals to pay for things they may not want so that others may benefit."

The further libertarian objection here is an important and intuitively appealing one, and we should reject it only with good reason. If the objection were indeed compelling, what would we be looking for in non-excusable public goods which would suggest that because of their very nature, we cannot coerce individuals into contribute toward their provision? Just what is the moral content of non-excusability? One prominent candidate might be this: non-excusable goods can't meaningfully be said to be capable of *acceptance* in the way merely non-excludable goods (or, for that matter, any other goods) can be described. We can *receive* non-excusable goods, but the fact that we play no significant part in determining whether or not we will receive them once they are provided renders them incapable of being *accepted* by a moral agent. And things which cannot be genuinely accepted, the libertarian would beckon us to conclude from this, cannot entail obligations to pay.

The distinction between the acceptance and the receipt of a good is articulated quite effectively by A. John Simmons (1979:Chapter 5) in his discussion of the Hart/Rawls Principal of Fair Play (what we have called here the *fairness principle* or more simple the "principal of fairness"). Simmons' primary goal is to argue that non-excludable (what he refers to as "readily available") goods are more commonly "accepted," while non-excusable (what he refers to as "open") goods tend, for the most part, toward being merely "received." The thrust of his discussion, however, is to suggest that the correspondence is not one-to-one, and to lay out the conditions under which any type of good—even an "open," non-excusable good, may in fact be *accepted.* Simmons enumerates a number of criteria, among which four essential ones can be distilled for our purposes. To *accept* a public good, one must:

i. know that the good is the product of a cooperative venture,
ii. desire the good,
iii. make an effort to acquire the good,
iv. not make an effort to avoid the good.

We can gather from Simmons' discussion that (i) and (ii) in combination with either (iii) or (iv) would be sufficient to indicate that one has *accepted* (and not merely *received*) a particular good. Were this possible of non-excusable goods, we might have a serious challenge to the argument from non-excusability (II), to the extent that it would demonstrate moral agents can make an active decision as to whether and how they will "receive" an ostensibly "non-excusable" good—in effect, *accepting* it. Perhaps just as important, an adherent to the argument from non-excusability could effectively defeat these criteria as too inclusive by demonstrating that they too readily mark as "accepted" a non-excusable good that we might otherwise believe to be incapable of "acceptance." Can the application of Simmons' criteria in fact challenge the argument from non-excusability (II), or are they indeed "too loose" and prone to too eagerly identifying *acceptance* of a good?

Let's step back a moment from the particularities of various environmental goods and take as an example the traditional textbook case of a non-excusable public good: national defense. This appears to be a unambiguous instance of a good which, once provided to a group of persons, cannot be denied to any member regardless of whether she contributed toward its provision (i.e., is non-excludable) but also cannot be avoided by any member regardless of her disposition toward it (i.e., is non-*excus*able). Is it possible that the case of national defense provide us with a counterexample of how these criteria might mis-specify the *acceptance* of a non-excusable good?

Note that the first three criteria, (i) through (iii), do not by themselves or in tandem appear to be terribly decisive for our example. Knowing that national defense is provided by virtue of the cooperative efforts of others (i) is something that could be described of just about anyone.[4] Similarly, condition (ii) cannot make or break the case for *acceptance*, since the adherent to the argument from non-excusability (II) denies that the desires and of agents involved are relevant, given the non-excusable nature of the good alone. Moreover, for criterion (iii) to wrongly attribute acceptance of a seemingly non-excusable good such as national defense, we would have to imagine a situation where a "beneficiary" of the good left the inaccurate impression that she was acting to acquire the good when in fact she wasn't. Such situations are difficult to imagine for any good we might consider "non-excusable," not the least of which national defense. Were one to make an active "effort to acquire" benefits from a nation's military structure—say by sending the local military installation photographs of everything around one's home that one would like "specially protected" in the event of a national emergency—there's not much room for confusion over whether one is "accepting" national defense or for misapprehension of one's dispositions about it. We wouldn't, of course, anticipate such actions even from those who really do "accept" national defense and the cooperative schemes which underwrite it (taxes and voluntary military service), much less from those who are genuinely indifferent to it.

It appears that, if Simmons' criteria for the "acceptance" of a good are to be at all decisive in helping us navigate the moral status of the argument from non-excusability, our attention must rest upon criterion (iv). If, in other words, the receipt/acceptance distinction is to break down in describing non-excusable public goods (and the advocate of the argument from non-excusability will hope that it does, since she wants to deny that such goods could ever be "accepted"), it must do so here. Is it possible for this criterion to misfire, to falsely deem the mere receipt of a particular good "acceptance"? Unlike its counterparts in (i) to (iii), it does seem plausible that one's not making an effort to avoid a non-excusable public good could be misinterpreted as "acceptance" of the good. In the terms of our example here, if one were genuinely indifferent to national defense, one's failure to make some effort to avoid receiving the good need not signal one's "acceptance" of the good. One could conceivably resign oneself to receiving the good in spite of one's reluctance, and behave no differently from those who relish the good and may truly be said to "accept" it. There are, in other words, limits to what we might conclude from one's relative inaction in such cases, and to hinge "acceptance" of a non-excusable good on one's failure to act to avoid the good—regardless of how one stacks up on the other three criteria—indeed suggests that we will be putting more goods into the "accepted" pile than we really should, and may risk wrongly subjecting some individuals to the responsibilities entailed by this presumed acceptance.

Note, however, that what we are considering in weighing criterion (iv) are the behaviors of those "consuming" the good, not—as the adherent to the argument from non-excusability would ask of us—the nature of the good itself. What this suggests is that a good's "non-excusability" is less a discrete characteristic of the good than a continuum on which we may place all publicly provided goods. Recall that previously we noted that a public good's relative non-*exclud*ability rests less on something inherent in the good than in the economics of the cooperative scheme which provides it—that is, a good is a "public" one to the extent that it is not cost-effective to produce the good *and* to exclude non-contributors. Something similar might be said of goods which can be described as non-excusable (although we begin to stretch the imagination a bit in doing so). We could argue, for example, that one could ultimately "excuse" oneself from even a national defense scheme: citizens who would prefer *not* to enjoy its benefits or contribute toward its provision or may submit to potential enemies of the state their names and addresses, so that they may be the first ones rounded up in the event of an invasion. Similarly, non-contributors can install floodlights to shine on their homes in the event of an aerial invasion, so that they may stand out more prominently as targets.[5] Such feats, however, do not involve simple efforts to *avoid* of the good, or even the mere alteration in lifestyle that Simmons (1979:30) would further include as within the bounds of our criterion (iv): they require that one go to great lengths to actively *escape* it. It is because we consider such an effort from people indifferent to the good *too demanding* in some instances that we do not consider its absence a sign of one's *acceptance* of the good. Indeed, we might consider it overly burdensome to ask an individual to take such steps to avoid the good. Such goods, we might deem, constitute genuinely non-excusable public goods, and suggest that we have defeated the ability of criteria (i) through (iv) to mark the genuine "acceptance" of a good otherwise thought to be non-excusable. The libertarian who touts the argument from non-excusability to justify withholding her contribution toward a public good can point out that some public goods are simply too pervasive to expect her to go considerably out of her way to escape consuming something to which she is indifferent.

Further specifying a non-excusable good in this way, however, lends no particular credence to the argument from non-excusability. In order for us to decide whether or not an effort to avoid a non-excusable good is "too demanding," we still need to make moral judgments about the effort expended, and what would be considered a "fair" amount to indicate that the individual expending it genuinely wished to avoid consuming the good in question. And this requires us to appeal to *something outside of the good itself*. Its non-excusability, in other words, is still not something given in the good, as the argument from non-excusability holds, but rather the product of a moral decision about whether one has acted sufficiently to avoid a good toward which one expresses indifference and an unwillingness to pay. Sometimes, the effort required to demonstrate this may be fairly mild; in other instances it may be fairly strenuous. In either case, however, the good is not intrinsically "non-excusable" but rather considered so on the basis of what we can reasonably demand of one who claims not to want it (much less contribute toward its provision) in order that they avoid benefitting from it. More important, I think it can be further seen that what we must appeal to in order to make such moral judgments are precisely the *preferences* that both arguments (I) and (III) above invoked and that the argument from non-excusability (II) considers irrelevant or, at best, secondary.

Another example might be useful here. Consider the following scenarios:

1. I scale the wall at an outdoor amphitheater in order to enjoy for free a concert that is provided to a paying audience
2. I walk into an outdoor amphitheater to enjoy a concert without paying the "suggested" donation solicited at the door and collected in basket on the "honor" system
3. An amphitheater is built such that my home on a nearby hillside has an unobstructed view of the stage

In none of these instances am I depriving others of anything for which they have paid (the good clearly exhibits jointness in consumption), yet our moral intuitions might tell us that relative excludability and excusability play a role in distinguishing the cases. We can, it seems, draw a line between the second and third scenario, such that while I may be considered to be acting unfairly in scenarios (1) and (2), this would not be so in the case of (3). Regardless of whether measures to exclude me in scenarios (1) and (2) would be cost-effective, it would seem wrong if not ridiculous in the third scenario for the owners of the amphitheater or the paying concert goers to "exclude" me, say, by asking that I step out of my house and go elsewhere during times of a concert, so that I can demonstrate a good-will effort to avoid consuming the good. Similarly, while it might seem morally defensible for them to exact payment from me in (1) and (2), this would seem to be patently indefensible in (3); I am in my own home, and the concerts in the amphitheater are goods to which I have access simply by virtue of my staying at home. The previous owner of the home could certainly have included the future positive externality of the concerts in the price of the home had she wanted to and had she known the amphitheater would be constructed (and she may well have, for all I know), but I don't seem to be under any obligation to the concertgoers to contribute toward the provision of the concerts in the amphitheater. Thus while in each scenario I have apparent unrestricted access to a good provided by the cooperative efforts of others yet for which I've not paid, scenario (3) seems morally different from the first two, and that difference seems to revolve around the fact that others have provided a good from which I am not easily *excused* and which I must make a tremendous effort to avoid consuming.[6]

Consider, however, yet another scenario:

4. I move into a home precisely because I am aware that it overlooks an amphitheater which hosts concerts that I know I will enjoy

Though paying concertgoers at any particular performance might concede that the concert is no less "excusable" to me than those in scenario (3), scenario (4) might strike us as somehow different. In (4), the amphitheater exists before I approach the situation, and I assume ownership of the home with the intent of listening to the concerts. What constitutes the difference between this case and that of (3)? Is it a cognitive one? There are undeniably differences in my knowledge here, though it is not clear that they are morally relevant differences. While it may be the case that I don't know prior to my purchasing the home I have in scenario (3) that I would later acquire the amenity of the nearby amphitheater, and that this amenity is known to me before my purchasing the home in (4), in each scenario I have the same understanding of the good provided—something from which I benefit without making a contribution to its costs. The fact that I didn't know of the amphitheater ahead of time in (3) seems less important than the fact that I *was* capable of knowing that *if* one were to be built within view of my home, I would benefit from it. And given what I probably already know about amphitheaters and concerts, it would have been fairly safe for me to assume that were one to be built, it would exist to serve patrons who pay an admission fee at the entrance, and that my enjoyment of the concerts would circumvent this system. In scenario (3), all of this may or may not have been a decisive factor in my purchase of the home; in scenario (4) it clearly was. Yet my knowledge of *how I might benefit from such a good* is the same in both cases. This doesn't entail that we are morally accountable for all counterfactual situations in which we might be non-excusably exposed to public goods we might enjoy and would have sought out anyway. The point is that the *cognitive* difference between entering a situation where a good is already so provided and one in which it can *potentially* be so provided is not one that alone can justify morally one's unpaid enjoyment of the good in the latter case. It is important that we not underestimate the weight of this point, since it is probably this intuition—that scenarios (3) and (4) are different because I know about the added amenity prior to my taking the house in the latter—on which we might initially rest in order to distinguish the two cases. While it may be true that the two scenarios involve differential *knowledge* of the immediate prospects for an amphitheater visible from my home, this is not a morally sufficient piece of knowledge here; nor is it the type of knowledge that Simmons is addressing in laying out the conditions for "acceptance" of a good. The knowledge that *does* matter morally—the knowledge that I *would* benefit without contribution from such an amphitheater regardless of whether one actually existed on that spot—is common to me in both scenarios.

Is the apparent difference between (3) and (4) then an *affective* one? Again, this doesn't seem to be the case. My desire for the concerts need not be any less in (3) than in (4). Certainly, I may also be indifferent to the concerts in (3), and this is clearly not the case in (4). But how would this potential difference in my desires manifest itself? Would we anticipate that I leave my home during concerts in the former scenario, and not in the latter? As we've already noted, this would seem to be an unreasonable expectation of me; we wouldn't anticipate one going to such lengths to *avoid* something which one wouldn't pay to embrace. Yet it would seem as unreasonable an expectation in scenario (4) that I refrain from purchasing the house altogether, for fear that I might enjoy the concerts. For all the concertgoers know, I might be just as indifferent to the concerts in (4) as I am in (3). The efforts I might make to avoid the good made "inexcusably" available to me, in other words, are no more likely in the former case than in the latter, and need not betray my genuine desires. Thus the presence or absence of such actions don't seem to help distinguish (3) and (4). The fact that I don't go to great lengths to abstain from the good in (4) signals my "acceptance" of the good no more than it does in (3).

We thus appear to have a problem: I have, in scenario (4), all of the seeming characteristics of what we might normally call a "free rider"—that is, I actively consume a good I know to be provided by the collective efforts of others but toward the provision of which I do not contribute. Yet there do not appear to be any morally relevant cognitive or affective differences between my actions in (4) and those in (3), where it is unlikely than anyone could consider me a free rider. Suggesting that I "accept" the good in scenario (4) while I merely "receive" it in (3) doesn't really tell us anything here unless we can point to such differences. Thus if we say that I'm acting within the bounds of fairness in enjoying the concerts without paying for them in (3), we lose any grounds for condemning my actions in (4). Accordingly, if we find grounds for condemning my actions as "free riding" in (4), we lose the ability to exonerate my actions in (3). Whatever moral wedge we might wish to drive between the two cases, it doesn't appear to rest on the "acceptance" or "receipt" of the good. If the two scenarios are morally different, something else must be distinguishing them.[7]

This point has important implications for non-contributors to the provision of public goods. If our only defense against compulsory contributions for non-voluntary benefits rests in the claim that the good is non-excusable and therefore cannot be "accepted," then we don't appear to have a sufficient rationale for withholding our contribution. To argue that such a good is non-excusable and incapable of *acceptance* would beg the further question of what would be morally required to signal "acceptance." The cognitive and affective criteria of the sort suggested by Simmons do not answer this question; they merely reframe it. Much like the relevant question for exc*lud*ability is not whether non-contributors can be excluded *simpliciter*, but rather whether or not it is practical or cost-effective for contributors to exclude them, the relevant question for exc*us*ability is not whether I can simply *excuse* myself from the enjoyment of the good provided via the contributions of others but rather *how much* and *what kind of* effort is morally required of me to do so in order for me to portray myself as an "honest holdout." The examples above do not suggest that an adequate distinction between the *acceptance* and *receipt* of a public good does not exist, but rather that if one does, it must rest on a moral difference in the consumer's *actions* or *preferences* and not the good itself or the scheme which confers it. Thus to suggest "compulsory contributions to non-excusable goods are morally wrong because the goods cannot be *accepted*" is to argue circularly: what would make such compulsion wrong depends entirely on the moral content of what we wish to consider the "acceptance" of a good.

"Non-excusability," in short, and in spite of the superficial characteristics suggested by examples like national defense, appears to be less a matter of the ubiquity of the good in question than of some negotiated congruence between its provision and the preferences of those consuming it. We might hope that those preferences are betrayed by the relative efforts one makes to enjoy or avoid the good, but such efforts cannot themselves be used as non-normative markers of one's desires, since they too are subject to moral scrutiny. If there is something different in the character of the free rider and of the honest holdout—something of moral significance which would suggest the latter can withhold his contribution with impunity while the former cannot—we apparently need to look further, perhaps at those preferences themselves.

NOTES

1. Richard Arneson (1982:62–63) discusses two different classes of individuals involved in the provision (or non-provision) of collective goods: the *nervous*

contributor and the *reluctant contributor*. The former, Arneson suggests, will contribute only if she can be assured there will be a sufficient number of contributors to actually secure the good; the latter wishes to contribute his fair share toward the provision of the good, but will do so only under the assurance that others will contribute their fair share. This appears to be a largely psychological distinction, and the two would likely be all but indistinguishable in practice: the former will withhold her contribution until she can be assured it will make a difference in obtaining the good, the latter will withhold his contribution until he can be assured he won't be a sucker. Moreover, given the earlier arguments in Part I, we might surmise that the lack of information that undermines such collective dilemmas is likely to ensure that both the nervous contributor and the reluctant contributor remain, for all intents and purposes, non-contributors. Schmidtz (1991:66–78) argues that the uncertainty about the cooperation of others that plagues prisoner's dilemmas of all sizes can be mitigated via *assurance contracts*, wherein a contracting cooperator secures the right to a "refund" of his contribution if a sufficient number of other participants is not achieved to secure the good (or if his share of the secured good does not exceed the costs of his contribution). While assurance contracts have a ring of plausibility in a number of situations, it's not clear that such contracts would provide an additional incentive in cooperative ventures necessary to secure large-scale environmental goods, especially given, again, as we've seen earlier in Part I, the potentially catastrophic effects of non-cooperation. If one were already disinclined to contribute toward averting the famine, disease, etc., that would likely accompany severe global warming, would she be any more likely to contribute once assured that if others are similarly inclined and it's not averted, she can get a full refund?

2. For the sake of clarity, it might help to emphasize that this characterization must describe non-excusable goods as a subset of non-excludable goods to be coherent, such that, though all non-ex*cus*able goods are also non-ex*clud*able, it is *not* the case that all non-ex*clud*able goods meet the conditions of non-ex*cus*ability. While it is possible, in other words, to describe a good which individuals must be able to consume *if they try* but which they may avoid if they don't, it would be nonsensical to describe a good which individuals *must* consume *simpliciter* but from which they reasonably can be excluded. The weight of the phrase "if they try" and the coherence it lends to the moral distinction required here is precisely what is at issue in this chapter.

3. Given these examples, we might suppose that, unlike the larger category of non-excludable goods, all non-excusable goods must also exhibit jointness. This certainly sounds plausible, and empirical examination might well reveal that the proportion of non-excusable goods which are jointly consumed is higher than the proportion of jointly consumed goods among the more general category of non-excludable goods. It's important to note, however, that this is a matter of contingency and not of logical necessity. Suppose ten people are sealed in a vault for a finite period with a limited supply of oxygen. If they cooperate and take turns holding their breath for short periods, the oxygen supply would be sufficient to keep them alive until the vault opens. Oxygen becomes a non-excludable, non-excusable public good to them (albeit one that is better described as "reproduced" among a very small "public"). No individual can feasibly be prevented from consuming the sustained oxygen which

the cooperative scheme would provide (short of killing him or her), yet there are no means for any individual of opting out of the cooperative scheme, since one is compelled to breath the available oxygen in the vault, the supply of which is made more abundant by virtue of the cooperation of others. The oxygen, however, is not in joint supply; unlike the oxygen outside the vault, one person's consumption of it affects the amount available for others to consume. Indeed, the good's availability to others is inversely proportional to the amount consumed by any individual. This example, however implausible, should be sufficient to demonstrate that jointness in consumption is no more an inherent feature of non-excusable goods than it is an inherent feature of public goods in general.

4. It's not entirely clear what type of situation Simmons has in mind for which "acceptance" of a good (as opposed to merely "receipt") hinges on this cognitive component. Few public goods—non-excusable or otherwise—come our way without our knowing that it is some sort of cooperative scheme which has provided them.

5. Such examples are along the lines of Nozick's (1974:94) invitation to imagine oneself refusing to participate in a cooperative street-sweeping venture in one's community: "Must you imagine dirt as you traverse the street, so as not to benefit as a free rider?" While Nozick is using "free rider" here somewhat more generically than in the current discussion, the point remains similar.

6. This point would be more abundantly evident had it not been the case that I actually enjoyed the concerts. They could just as easily have consisted in music I didn't enjoy—or even found annoying—thus constituting what I would consider to be a negative rather than positive externality. In this case, it would seem clearly wrong to impose an onerous burden upon me and then to charge me for it.

7. This seems to be a clear example of what Shelly Kagan (1988) describes as a *contrast argument*. In this instance, we hold all other morally relevant conditions constant and vary the one in question to show that, since the moral status of the contrast cases doesn't change significantly, the factor in question must not be morally relevant. Given this, all necessary restrictions and cautions should apply to this conclusion, to the extent that the additivity of the factors involved must be assumed for the argument to be coherent. At most, we are able to say that for examples meeting identical conditions, we would encounter the same difficulty in distinguishing the act of acceptance. We will see shortly that Kagan's account of the "Additive Fallacy" will be relevant in calling into question the argument from preferences.

Chapter 6

Intentions, Free Riders, and Honest Holdouts

We've seen thus far that it is very difficult, if at all possible, to maintain arguments from non-excusability (II), as I described them in the last chapter. Try as we might, we must appeal to something outside of the non-excusable nature of the good to distinguish morally, on the one hand, the honest holdout who genuinely prefers not to consume the good and certainly not to contribute to its provision but who cannot easily avoid consuming it and, on the other hand, the free rider who similarly withholds his voluntary contribution but actually desires the good (we must, for example, make a moral decision as what sorts of effort would be sufficient to count as a genuine attempt to avoid consuming the good). In terms of environmental politics, this suggests that even those individuals claiming to be unconcerned about the effects of climate change and other ecological maladies and unwilling to contribute voluntarily to their prevention might still justifiably be coerced into contributing in some way. If the libertarian intuition that we cannot be made to pay for something we don't actively solicit or purchase and cannot even avoid "consuming" has a solid moral foundation, we haven't found it yet.

Arguments from preference (I), also described in the last chapter, has merit insofar as both the environmentalist and the libertarian could be said to assent to them in principle: the free rider and the honest holdout have different desires and preferences—one seemingly pernicious, the other apparently sincere—and it would be ideal to treat them as morally distinct in accordance with those desires and preferences: the free rider prefers to consume the good and the honest holdout is genuinely indifferent, so common sense would instruct us to exact a contribution from the former but not the latter. The problem, as we saw, is that those preferences alone are all but indistinguishable when exercised, and there are unavoidable epistemic limits in genuinely identifying genuine honest holdouts when everyone who withholds a voluntary contribution toward the provision of a public good such as stable planetary temperatures has reason to claim he is an honest holdout and not a free rider.

Focusing on *preferences* alone in any case stacks the deck in favor of the free-market libertarian. While it is conceivable that what motivates us to purchase purely "private" goods on a market such as a jar of jam or a greeting card is a "preference" for having that item to not having it, public goods, from what we've seen so far, are not so readily disentangled from larger social practices. A "preference" for environmental quality does not translate into a simple market purchase; it must be coordinated with many people, and we must form some expectations about what those people will do as well as intentions about what we will do given our preferences and our expectations. It is, arguably, these intentions that retain moral content when it comes to assessing the various actors involved in the provision of public goods.

What might serve as a viable candidate for an intention which could potentially distinguish the free rider from the honest holdout? Consider one that seems non-controversially attributable to the typical free rider:

(i) To consume a benefit he knows to be the product of the cooperative efforts of others

Note that the intention includes a cognitive component; it is, in fact, the first (i) of the four criteria we distilled from Simmons (1979:Chapter 5) and examined in the last chapter. But it attaches to the awareness of the cooperative nature in the source of the good's provision a definite intention to actually "consume" it, however this might be done in the case of the environmental "goods" that are our principal concern here. Cullity (1995) considers this intention in asking much the same question we are asking. He suggests, quite rightly it seems, that it does present at least a *prima facie* distinction between the free rider and what we are calling the *honest holdout*.[1] While both might be well aware of the good's origins in a collective, cooperative scheme, only the free rider acts with the intention of consuming the good. The honest holdout may consume the good no less than the free rider, but it is not his *intention* to do so.

Yet while this intention may uniquely characterize the free rider in comparison to the honest holdout, it is by no means clear that it contributes anything to what we might consider the morally objectionable content in the free rider's actions. Note—as does Cullity—that the same intention can be said to be held by the cooperator who contributes to the benefit-conferring scheme. She too intends to consume a benefit she knows to be the product of the cooperative efforts of others. She too knows that others will have to bear certain costs in order that she may enjoy the good. She is aware of this when she consumes the good. Yet she, unlike the free rider, willingly contributes a share of the burden herself, and thus situates herself in the realm of what we might consider to be "fair." The intention thus, as Cullity puts it, "under-specifies" any distinction between the free rider and the honest holdout, to the extent that those who cooperate in the provision of the good share the intention no less than free riders. If we are to have an intention-based moral distinction between the free rider from the honest holdout, it will need to be one which also distinguishes the free rider from the cooperator. Intention (i) does not appear to do both.

Suppose, then, we consider a second intention which seems unambiguously true of the free rider:

(ii) To withhold contribution to the cooperative scheme

This appears to clearly distinguish free riders from cooperators; while both consume the good they know to be the product of a cooperative effort, the former is a "free rider" precisely because he intends to consume it without shouldering a share of the burden in the good's provision. Unfortunately, with (ii), we lose the ability to draw the distinction which was our initial goal: that between the free rider and the honest holdout. While the latter may not intend to consume the good, he intends no less than the free rider to withhold any contribution to the good's provision. With (ii), we thus draw a prominent line between the free rider and the cooperator, yet we are unable to employ it in drawing a similar line between the free rider and the honest holdout, and we are no closer to our goal.

Why, we might ask, take the two intentions in isolation? It seems as if in tandem the two might adequately distinguish the free rider from his counterparts, and thus do the work we were hoping that they would do. Table 6.1 lays out the status of both intentions for each agent.

As we can see, the free rider shares intention (i) with the cooperator, and shares intention (ii) with the honest holdout, but he remains of the three

Table 6.1 Intentions (i) and (ii) of Cooperator, Free Rider, and Honest Holdout

Intention	*Cooperator*	*Free Rider*	*Honest Holdout*
(i) To consume benefit or good provided by cooperative scheme	✓	✓	
(ii) To withhold contribution to the benefit or good		✓	✓

agents the only one of whom we can say he harbors *both* intentions (i) and (ii). Conversely, we might say that what makes the cooperator and the honest holdout morally unique—and thus differently culpable from the free rider—is the fact that they have only one of the two intentions, a state which is morally benign in comparison to that of the free rider. If this is so, perhaps the argument from preferences (I) that we visited in the last chapter—with *preferences* now understood more robustly as involving expectations and intentions as to how one will in fact *act* on one's preferences—has some practical applicability after all. Does this seem plausible?

Two concerns—both of which traceable to the immeasurable contributions of Ludwig Wittgenstein's later work to modern philosophy[2]—should give us pause before we conclude that we've resuscitated the class of *arguments from preferences* for practical application to situations such as the formulation of environmental policy. Foremost is the fact that we are treating mental states as if they were readily graspable entities existing on their own and capable of manipulation in much the same way we manipulate hammers, nails, and wooden planks or, in this case, numbers on paper. Moral philosophers unpersuaded by Wittgenstein's legacy might be forgiven for investing things such as expectations and intentions with an ontological status that permits such properties as "additivity," but neither ordinary language nor a more skeptical moral philosophy finds this an easy task. Even moral philosophers not expressly troubled by these ontological implications may find it a bit odd to attempt to "sum" the moral characteristics they examine.[3] If we are searching for a moral distinction based on the relative mental states of the cooperator, the free rider, and the honest holdout, the mere fact that *we* can "add" two separate intentions that we attribute to them in some combination does not by itself map to some corresponding additive process in the minds of those individuals.

Second, and very much related, mental states such as *intentions* do not exist in a vacuum which permits them to retain meaning in and of themselves. They make sense to us only in the context of a common social and linguistic practice—what Wittgenstein famously labeled *language games*—and we virtually assure misinterpretation of them when we attempt to attribute meaning to them outside of those practices. We must, in other words, know something about the cultural, social, and institutional setting in which preferences are formed, what options these factors make available, and what the agent could reasonably expect as a consequence of his actions in such contexts in order to attribute meaning to his intentions to act in particular ways—certainly to pronounce those intentions morally sound or suspect. That he has them is not terribly informative in and of itself. As Wittgenstein would so aptly put it, we have to understand the "rules of the game" that provide meaning to the participants before we can play the game ourselves.

To clarify this point, consider a different example in which two seemingly innocent intentions appear to sum to something morally objectionable. Snake is a bank robber. Snake approaches a bank with two intentions in mind: (i') to enter the bank and request money that he has himself never deposited; and (ii') to leave the bank carrying more cash than he had when he entered. Since we already, for various reasons, understand bank robbing to be morally wrong, we might think that it is the combination of these two intentions which marks the bank robber as blameworthy. Yet one might readily point out that Snake shares intention (i') with any loan applicant who enters the bank, and (ii') with any depositor who wishes to withdraw cash from her savings account. We wouldn't, however, want to argue that since Snake is indistinguishable from the loan applicant on intention (i') and from the depositor on (ii'') that his act of robbing a bank is morally acceptable. But why is this? Something is different about Snake's moral status, insofar as robbing banks is different from applying for loans or withdrawing cash from one's savings account, but what constitutes his culpability here if not the combination of these two singularly benign intentions? In this case, it seems to be something lying outside of Snake's intentions to demand money and leave the bank with it: the existence of rules which properly allocate funds to loan applicants and depositors wishing to withdraw cash from their accounts. In order for us to distinguish a loan application from a stick-up, we need to be able to stipulate the conditions under which one may request a loan, the conditions for repayment, the collateral required, etc. These things are independent of the intentions of any particular loan applicant. One might well object that the loan applicant is typically different from the bank robber in that he has the further intention of paying back the money being demanded. This is, however, not a necessary condition for a loan applicant; one could borrow money from a bank with no intention of repaying the loan. For that matter, Snake could hold up a bank at gunpoint and promise the teller that he'll be repaying the bank in full and with interest once his pick in the next horse race is a winner; this would rightfully go down on the police blotter as no less a hold-up than one where no such promise is made. What remains the distinguishing characteristic here is the set of rules—legal, moral, conventional, etc.— regarding the proper application for a loan which both the loan applicant and the bank robber encounter upon entering the bank but which the latter violates by brandishing a gun and threatening a teller. This would be a different situation from a loan application even if he had every intention of repaying the money taken.[4]

Much the same could be said for Snake's differences with the depositor withdrawing cash. It is not just the fact that he intends both to demand money and to leave the bank with it that distinguishes him from the depositor who has no intention of asking for money to which she has no claim. It is rather the fact that the depositor *possesses* the money she is retrieving from the bank; she has a claim to it, by virtue of a legal system which stipulates the conditions of ownership, which Snake can't say of the cash he will be leaving the bank with. Again, these conditions and rules appear to be independent of the intentions of both Snake and the depositor. One might object that we can say this is nothing but a more precise specification of intentions: the depositor intends to leave the bank with cash to which she has a legal claim, while Snake intends to leave the bank with cash to which he does not have a legal claim. But this is merely to invoke intention (i') in regard to Snake, and to beg the further question of where such ownership claims come from. We are far from being able to conclude that Snake can be distinguished from the loan applicant and the depositor on the basis of intentions alone.

Do such intention-independent conditions exist in the provision of non-excusable public goods? This seems clearly to be the case when we articulate the distinction between the free rider and the cooperator. Neither the principle of fairness nor, by extension, the argument from non-excludability which deem the actions of the free rider "unfair" in contrast to the cooperator depend for moral weight on the intentions of either party. It is, as we've seen, the fact that one *could* "get something for nothing" in regard to non-excludable public goods—regardless of one's stated or unstated preferences—that operate in the argument from non-excludability. We were aware of this before we even laid out intentions (i) and (ii). Can the same be said of the main distinction of interest here, that between the free rider and the honest holdout? Unless the adherent to the argument from preferences can demonstrate that no intention-independent moral factors are required in the distinction between the free rider and the honest holdout, such arguments are in trouble. In attempting to single out the intentions of the free rider as morally objectionable, the burden of evidence is on those adhering to the argument from intentions to discount competing, intention-independent moral factors. And given that it seems clear that the distinction between the free rider and the cooperator does depend on such factors to render intentions (i) and (ii) of unique moral significance in their combination, it seems a reasonable assumption that the same might be true if those same combined assumptions are to somehow distinguish the free rider from the honest holdout as well.

Suppose, however, that the additional moral factors necessary were themselves further intentions. Would the argument from intentions then remain intact? We might take Cullity's cue and consider the possibility that the free rider and the honest holdout may be distinguished not solely in terms of their intentions about *their own actions* but rather in terms of their intentions about *the actions of others* (Cullity, 1995:26).[5] One of the more prominent candidates for such an intention might be:

(iii) That others *not* withhold their contribution to the benefit-conferring scheme.

The honest holdout, we might assume, is prepared to extend the privilege of non-contribution to everyone else. Since he is indifferent to consuming the good provided by the benefit-conferring scheme, he is not wedded to the condition that *others* intend to cooperate in order that the good be provided in spite of his unwillingness to cooperate. This is not true of the free rider. His intention to consume the good while withholding a contribution of his own to the cooperative scheme is matched further by the intention that *others do not take the same option* since, as we have seen, if k-1 of them did, he would not have the good to enjoy.

Table 6.2 includes (iii) in our list of intentions attributable to each agent.

Among the first things that might strike us about this intention is that it has the same characteristic we observed in intention (i), namely, that it is an intention which the cooperator too must hold. To the extent that the cooperator intends (i) to consume the benefit provided by the cooperative scheme, she must also intend that others contribute toward the provision of the good; intention (i) wouldn't make sense if she did not believe this. Yet if the cooperator holds this intention no less than the free rider, we must once again ask if we have really earmarked what is morally repugnant about the free rider.

The libertarian, however, no doubt trains his eyes upon the last column and the solitary intention of the honest holdout to bow out of the whole affair. "Take particulate matter out of the air you breath, if you must; pay a federal tax surcharge to fund intrastate high-speed rail transit; change your very lifestyle to lower your carbon footprint. Just leave me out of it. I elect not to participate in your 'cooperative' (really more *coercive*) plans, and I frankly don't care whether any of you do. Whether or not your aims are not achieved, so be it; I was indifferent to them in the first place." Surely, we might sympathize, the honest holdout has distinguished himself here insofar as, unlike the free rider, his lack of any intention to consume the public good provided by the cooperative actions of others is consistent with his indifference to whether anyone else contributes her cooperative efforts. He is unique in that he neither cares for the good nor whether it is ever provided and consumed by virtue of

Table 6.2 Intentions (i), (ii), and (iii) of Cooperator, Free Rider, and Honest Holdout

Intention	*Cooperator*	*Free Rider*	*Honest Holdout*
(i) To consume a benefit or good provided by cooperative scheme	✓	✓	
(ii) To withhold contribution to the benefit or good		✓	✓
(iii) That others *not* withhold contribution	✓	✓	

the efforts of others; his intention (ii) to withhold his "contribution" must thus be morally benign, mustn't it?

Despite a superficial, intuitive appeal in the libertarian's plea on behalf of the honest holdout, I don't think it does as much to *morally* distinguish the holdout from the free rider as it does to psychologically delineate them. Recall that we've previously denied the libertarian the appeal to arguments from non-excusability (II) since, as we saw, the non-excusable nature of the good alone cannot justifiably exempt from cooperative duties anyone who wishes not to pay for a good she is "compelled" to enjoy. What I am suggesting here is that neither does the libertarian have recourse to arguments from preferences (I); that is, the libertarian cannot deem the honest holdout less culpable (and thus exempt from contribution) than the free rider based solely on an appeal to the preferences and associated intentions of the relevant agents. Preferences and intentions are by themselves, at best, mental states or psychological constructs. They don't carry moral weight outside a system of rules of grammar, logic, or law that infuses them with moral meaning.

Suppose, then, that the libertarian does indeed offer just such a moral basis for identifying the honest holdout's intentions as distinct from the free rider's: between the two of them, it is only the honest holdout's intentions that are morally *consistent* on Kantian grounds of universalizability: the free rider alone holds all three intentions (i), (ii) and (iii) and thus is the only one among the agents here who cannot rationally will that others do what he intends to do himself; indeed, he must, as a free rider, hold both (ii) and (iii) and thus intend that others do precisely the opposite of what he intends to do. While both the cooperator and the honest holdout share one or two if these individual intentions with the free rider, Kantian universalizability of their intentions is possible: what each of them wills for herself she can rationally will for others. It is in this sense, one might suggest, that we may single out the free rider morally on the basis of his intentions.

Does this salvage the argument from preferences (and their associated intentions) as vindicating the honest holdout and exempting him from contribution? Not necessarily. What we are calling the "honest holdout" no less than the free rider may experience a contradiction between what he wills for himself and what he wills for others to the extent that intention (iii) actually characterizes him as well. Cullity's account is once again relevant here. The holdout in Cullity's primary example—the "Recalcitrant Fisherman" who claims indifference the cooperative efforts of his fellow fishermen to clean the lake that provides them with their livelihood—Cullity argues, cannot in sincerity intend that others follow suit in his refusal to contribute to the effort. His very livelihood depends on clean, inhabitable water for fish, and for him to not will that others reject (ii) and contribute toward the cleaning of the lake would be as irrational as the free rider willing that others withhold their contributions. While the free rider can sincerely admit that he intends others to avoid the route he takes and to contribute toward the good, the honest holdout's denial of this same intention is either disingenuous or incoherent, since he too depends on the cooperation of others for the good—substantively, if not consciously. Thus while it might initially appear that only the free rider is caught in a Kantian bind of being unable to will that his own actions be universally adopted, Cullity's example suggests that the honest holdout must also hold intention (iii), and that this subjects him to the same conflict that the free rider experiences between what he wills of himself and what he wills of others.

Surely, though, the diligent libertarian will not resign in defeat at this example; whether or not our honest holdout's stated indifference to the intention of others to contribute toward the good (iii) can be considered sincere or coherent must depend at least somewhat on the good in question. Suppose the good made available (and more or less non-excusable) to the Recalcitrant Fisherman by collective contributions was not a clean lake which would ensure adequate fish stocks but rather a series of decorative buoys on the lake intended merely to indicate points of varying water depth. We might describe this as something from which the Recalcitrant Fisherman would "benefit," insofar as fish seem to care about where they swim, and we might thus think it useful for the Recalcitrant Fisherman to know the depth of the water beneath him. He might even himself assent that he is better off knowing how deep the water beneath him is than he is not knowing this. Can we say, however, that it would be irrational for him to be indifferent to whether other individuals who earn their living by fishing on the lake contribute toward the decorative buoys? His and their livelihood do not seem to be riding on the provision of the good in the same way cleaning the water which threatened their ongoing stock of fish did. Surely, he would receive some "benefit" if enough of the others decided to contribute toward the good in spite of his recalcitrance, but there seems to be no irrationality or insincerity in his willing that these others do as he does and withhold their contributions. Unlike the cleanup of the lake's water, moreover, knowledge of varying points of depth in the lake is something he could, if he chose, provide for himself on a private market: he could purchase one of a number of electronic sonar devices to equip his own boat, however "non-excusable" knowledge of relative depths might be if provided by the decorative buoys. Thus while the good in Cullity's original example—a clean, inhabitable lake—might strike us as "vital" and "important" enough to the Recalcitrant Fisherman that his indifference as to whether others similarly withhold their contribution to the cleanup efforts would be insincere or irrational, there are a host of other goods which might be provided to him or to any honest holdout which are both more or less non-excusable and fully capable of rational and sincere indifference to both consuming the good and to the contributions of others. While intentions alone might appear insufficient to distinguish Cullity's Recalcitrant Fisherman and Fare Evader, they might well be sufficient in these latter instances.

Note, however, that we are right back to where we were with arguments from preferences. If mental states such as preferences and intentions are sufficient only under certain conditions which themselves are not contingent upon preferences and intentions, we would seem to have maintained the case that the argument is not a viable one. It must again appeal to something outside of the respective intentions of the free rider and the honest holdout in order to draw a moral distinction in their actions that would presumably vindicate those of the latter. What determines whether or not the honest holdout can rationally will that others withhold their contribution to the benefit-conferring scheme is not something he himself wills or prefers or intends but rather something about the good, namely, its relative importance in his life to his ongoing existence, well-being, etc. Some goods may be made available to him—goods which he may find difficult to avoid—which are clearly expendable in his life and may even be considered "trivial." It would not seem irrational for him to say of these goods: "Look, I really don't want or need this, and I am consuming it only because I would have to go to great effort not to; hence I'm not going to contribute to your collective efforts to achieve it on my behalf and on your own behalf, and I wish you would all consider adopting the same policy." Other goods may be provided by similar benefit-conferring schemes which might strike us as more critical to his ongoing well-being. Were the honest holdout to offer the same rationale for non-contribution as above, we might think it insincere—in that we would believe him to be strategically underestimating his stated preferences in order to shirk his burden in contributing to it—or we might think that he genuinely doesn't know what is best for himself. Thus in the case of the "trivial" non-excludable good, there seems to be a moral difference between the honest holdout and the free rider that rests largely if not entirely on their respective intentions: while consumption of the good is more or less compulsory for each, the free rider really does prefer having the good to not having it, intends to consume it, intends to withhold his contribution to its provision, and hopes that others don't follow his actions. The honest holdout in this case could, for all he cares, take or leave the good, and might rightfully claim to be unbound by the obligations to the benefit-conferring scheme. He is, if nothing else, coherent in his preferences on Kantian grounds of universalizability. In the case of the "non-trivial" non-excludable good, this difference is significantly harder to see. The free rider would clearly be insincere in disavowing the good, but we would have to believe something like this is true for one who purports to be an "honest holdout." Even if the good were more or less non-excusable, we would thus be unable to tell the two apart, since the "non-triviality" of the good would suggest that no rational person would go to any effort to avoid it.

There is one final very important objection that the libertarian would surely raise here, and it is one that we must consider. "Who are we," the libertarian might rightly ask, "to decide what should be considered 'trivial' and 'non-trivial' by the honest holdout? What it is worth to him, it is worth to him, and whether individuals can be compelled to pay more for something than it is worth to them is precisely what is at issue here. The overriding question in Part I has been simply this: Is there is anything that can shake our libertarian intuition that it is not permissible to coerce people into contributing toward public goods which they cannot readily avoid? To argue that the free rider may well be compelled to contribute toward non-excludable goods, that the difference between non-excludable and non-excusable goods doesn't help us distinguish the honest holdout from him, that the honest holdout *might* be morally distinguishable by virtue of his intentions, and only *then* hinge the leverage of these intentions on the relative 'triviality' of the good in question does nothing to establish an argument against this otherwise compelling libertarian intuition. If the good strikes the honest holdout as 'trivial' then this is all that matters, regardless of how important *we* might think it to his well-being, and regardless of whether we believe he is acting in his own best interest in being indifferent to it and to the cooperative efforts of others to provide it. To compel the honest holdout to contribute toward the good because *we* believe the good is 'non-trivial' enough to be in his interest in spite of his preferences is sheer paternalism."

There is indeed a decided lack of appreciation for this point in much of the literature on public goods—certainly environmental goods and the benefits bestowed by environmental policies—and the obligations which might oblige us to contribute toward their provision. Nozick, to his credit, had added to the Hart-Rawls *fairness principle* his concern that, *if* it were ever permissible to coerce an individual to contribute to a cooperative scheme, it would be necessary that the net benefits of the good to her must be greater than the net costs of her cooperation.

> At the very least one wants to build into the principle of fairness the condition that the benefits to a person from the actions of others are greater than the costs to him of doing his share. How are we to imagine this? Is the condition satisfied if you do enjoy the daily broadcasts over the PA system in your neighborhood but would prefer a day off hiking, rather than hearing these broadcasts all year? For you to be obligated to give up your day to broadcast mustn't it be true, at least, that there is nothing you could do with a day (with that day, with the increment in any other day by shifting some activities to that day) which you would prefer to hearing broadcasts for the year? If the only way to get the broadcasts was to spend the day participating in the arrangement, in order for the condition

> that the benefits outweigh the costs to be satisfied, you would have to be willing to spend it on the broadcasts rather than to gain *any* other available thing. (Nozick, 1974:94)

Even staunch defenders of the *fairness principle* would presumably assent to Nozick's concerns since, even for goods which might strike the "beneficiary" as "trivial," it might be permissible to coerce *some* contribution from her if its cost to her were even more trivial. The more important point here is the fact that Nozick clearly intends this assessment of costs and benefits to be a subjective matter from the standpoint of the individual deciding whether it is worth her while to contribute to the benefit-conferring scheme. When such "costs" and "benefits" are taken up elsewhere, they are often assumed to be purely non-subjective, and calculable strictly in terms of an impartial metric and without regard to the preferences—stated or unstated—of the individuals to whom they accrue.[6] George Klosko's (1992:Chapter 2) treatment of this issue remains a careful exception. In his extended discussion of the Principle of Fairness, Klosko draws attention to what he refers to as *presumptively beneficial goods*. While the label is derived from Rawls' account of *primary goods*—"things that every rational man is presumed to want" (1971:62)—Klosko intends presumptively beneficial goods to be a "much narrower class than primary goods" insofar as the former are "confined to public goods" and "must be indispensable to acceptable lives" (Klosko, 1992:40). He enumerates, by way of example, "physical security, comprised mainly of national defense and law and order, protection from a hostile environment, and provisions for satisfying basic bodily needs" (Klosko, 1992:40). We can no doubt include a number of environmental goods into this category, primarily those whose provision can be demonstrated to be critical to the ongoing viability of the planet and particular ecosystems within it.

Should the libertarian find any consolation in a formal account of *presumptively beneficial goods*? Without an argument as to what moral implications follow from them, it's not clear that she should. Klosko himself notes that he is merely appealing "to relatively uncontroversial aspects of our moral beliefs" (1992:40). This might strike us as agreeable enough in stipulating a class of goods that, regardless of one's beliefs, preferences, desires, etc., are indispensable to one's welfare. Whether any and all means of providing such goods are morally acceptable is not, however, as obvious. Just as a loaf of bread's "presumptively beneficial" status (if indeed we may extend Klosko's notion to private goods) would not, in itself, justify stealing it by one who truly needs it, neither should it alone permit us to impose bread upon another person and demand payment for it. There may well be circumstances under which stealing a presumptively beneficial private good is permissible, as well as those under which coercing a contribution toward a presumptively beneficial public good is permissible, but this permissibility cannot flow forth from the good's presumed benefits—not, at least, without some argument as to why we should believe this to be so. One can readily assent that a good non-excusably provided by a collective arrangement is indispensable to one's welfare, essential to one's physical security, fundamental to one's ongoing existence, etc. But one can also rationally compromise one's welfare, security, and ongoing existence if one perceives that the potential benefits of doing so outweigh the risks. Cleaner drinking water may well be demonstrably healthier and more conducive to a happier life than polluted drinking water, but Smithers may prefer to invest the money he is being asked to contribute toward its provision in something that he may believe to have equal or even greater impact on his well-being—perhaps even a different *presumptively beneficial* good. Provided he genuinely understands the potential benefits of cleaner water and the risks involved in spending his resources on alternative goods, he would appear to be acting prudently, if not "rationally," and our attempts to coerce a contribution for our preferred good on the grounds that he "doesn't know what is in his best interest" may rightly be considered paternalistic. It appears that, if we are to hinge the argument that the honest holdout cannot, like the free rider, be genuinely indifferent to the good and to whether others withhold their contribution to it *on the weight of what we believe it rational for him to do*, the argument will come up short. Even if our assumptions about "presumptive benefits" have merit, they are clearly not going to sneak past the diligent libertarian, as they probably shouldn't. If we are to indict the honest holdout along with the free rider, it won't do to argue that, because of the presumptively beneficial nature of the good, the former is not acting in his own best interest and should thus be compelled to do so for his own good.

This does not imply, however, that the argument from preferences (i) stands undiminished, for we still have another means of acknowledging that some public goods can reasonably be considered more "non-trivial" than others without paternalistic appeals to the welfare of their beneficiaries. If it won't do to raise the cognitive question of whether the agent recognizes the importance of the good to his own well-being, we might still raise the strategic question of whether he is sincere in his eschewal of the good, and whether his expressed indifference to whether others similarly withhold their contribution to the good is any more genuine than the free rider's expressed intention to do the same. We saw that free riders, by their very nature, must actually intend that others do not do as they do; whatever their claims, they must hope that others contribute to the provision of the good, lest k-1 potential contributors follow suit and the good remain unprovided. For some goods—goods which we may certainly wish to describe as "presumptively beneficial" and which may even be more or less unavoidable—we might draw a similar conclusion about those who claim to be honest holdouts. We might be able to say to them: "Look, this good that we are providing by our cooperative scheme is one that you cannot live without; it is absolutely essential to your well-being. Granted, it may not be our place to tell you that pursuing what is essential to your well-being is in your best interest, or to compel you to pay for goods that serve it. The problem is that, given the importance of this good to your ongoing existence, and given that we all agree that individuals tend to act on behalf of this, we simply cannot accept as sincere your claim that you really don't want the good. Decorative fishing buoys, neighborhood public address broadcasts, etc., however enjoyable we might find them, would certainly not strike us as goods critical to your welfare, and we'd understand it if you elected to bow out of the collective agreements which provide them—even if you couldn't help but 'enjoy' them. A stable global temperature, national defense, etc.—these things seem different. When you say that you are similarly indifferent to these, we have good reason to doubt that claim, if only because the *value* that *we* attribute to these things is so markedly higher than the value you claim to attribute to them. We are thus left in a situation where you are, for all intents and purposes, indistinguishable from the free rider who may make similar claims, but for whom these goods are important and worthy of contribution, albeit potentially attainable without that contribution. While there *may* be differences in your intentions in these cases, for all practical purposes, you and the free rider are on the same moral ground here: we see you as strategically underestimating the value of the good to you in order to shirk your share of the burden in providing it. If we may justifiably compel the free rider to contribute toward the good, we may justifiably compel you as well."

What we are really trying to distinguish morally here is not something in the character of non-contributors but rather something about the goods which they eschew (though, as we have seen, it is not the non-excusability *per se* of such goods) and the implications of their actions for those who do choose to cooperate. Neighborhood public address systems (Nozick), overnight shoe repairs (Cullity), etc., are not a sustainable environment, a workable and stable economy, or international security; not all benefit-conferring schemes, other moral intuitions quite rightly tell us, are equal. When my neighbors cooperate to decorate the city streets for a holiday, my failure to contribute here seems different than my failure to contribute when they cooperate to ensure a safe water supply. Even though the two benefits appear equally non-excusable, and even if the stated indifference of both the free rider and the honest holdout are somehow distinguishable, the moral weight which we might attach to the duty to cooperate in the former case does not seem to have the same mass as the duty in the latter. While the argument from preferences may tell us *something* about a subclass of non-contributors in the former, neither the argument from non-excusability nor the argument from intention is sufficient to single out a class of "honest holdouts" which are exempted from whatever compulsory efforts we may be justified in using to collect contributions. Arguments from non-excludability, it seems, prevail: if we have grounds for exacting contributions toward the good from free riders, we also have grounds for exacting contributions from those who profess indifference to the good itself and to whether others cooperate to provide it, since the epistemic limits inhering in the situation preclude us from drawing a meaningful, practical distinction between the two.

NOTES

1. Cullity is actually contrasting what he calls the "Fare Evader" and the "Recalcitrant Fisherman," but he is addressing more or less the same distinction. The two labels are derived from ongoing examples in his article: the former, of an individual who takes a ride on a public transit system but avoids paying the "honor system" fare, and the latter of a fisherman who opts out of a cooperative venture by other fishermen to diminish the pollution in a lake which has come to threaten fish populations and thus their livelihood. The "Fare Evader" seems clearly to be *free riding* on a non-excludable good (one that goes *un*excluded, at least, in this example), while the "Recalcitrant Fisherman" appears to be an *honest holdout* on a benefit that he could avoid only with great effort and expense to himself—roughly, one that is non-excusable.

It is interesting to note that Cullity's impetus for even taking up the question of a distinction between these two agents is quite different from the one here, and is in many ways antithetical to it. Cullity begins from the seemingly paradoxical notion that the free rider strikes us as acting unfairly yet, since he consumes a joint public good, is harming no one (recall that standard accounts of joint public goods assert that consumption by one person does not diminish the amount available to others). Cullity then goes on to ask what it is that is unfair in the free rider's actions, and sets up the Fare Evader and Recalcitrant Fisherman examples to put proper restrictions on how we might answer the question; if we assume the latter's actions to be morally defensible insofar as he has no choice but to accept the good, we would want to make sure that what we describe as "unfair" in the Fare Evader's actions should not similarly describe the Recalcitrant Fisherman's.

The initial assertion that the free rider is harming no one is, however, not so easily accepted and applicable only to a narrow range of *non-crowdable* public goods. As we saw earlier in Part I, for all individuals i_1 to i_k with an opportunity to consume a crowdable joint public good, there is an infinitesimal yet definite probability that their act of consumption will be the one to "tip the scales" and deplete the good. Thus while it is not the case that each free rider who withholds a contribution takes something for nothing, each does increment by a definite amount the probability that others will not benefit (though, since they also increment this probability that they too will gain no further benefit, I argued earlier that this is not a case where the demands of morality are at odds with the self-interest of any i_1 to i_k). All of this would be no less true for the honest holdout.

2. His posthumously published *Philosophical Investigations* is the most comprehensive in this regard.

3. See, for example, Kagan (1988).

4. By the same token, note that it is not the intention to threaten violence that distinguishes Snake from the loan applicant, since the latter can have every intention of filling out the necessary forms and eventually repaying the loan but also be fully prepared to take hostages if he is turned down. The fact that such threats are not expressed in most loan applications suggests less an inherent difference in the intentions of bank robbers and loan applicants than an institutionalized set of procedures which renders robbing banks unnecessary for most individuals.

5. Having *intentions* about the *actions of others* may sound a bit odd if we understand "intention" to refer to a sort of "motor" for strictly personal acts of volition: I *intend* to raise my arm, to invoke a classic example in action theory, and I do. How, one might wonder, can I *intend* that another person raise her arm? We should think, however, of "intentions" here more generally as encompassing not just one's one actions but the expectations we might have about their effects given what we anticipate others will do. These are in essence the sorts of intentions and actions social choice theory addresses, and what we having been dealing with throughout Part I (and, for that matter, will be contending with in Part II). Little of this is in fact at odds with ordinary language. For example, my *intention* to walk across a busy intersection at a green light is in some sense what gets my legs moving. But it is also an *intention* that drivers traversing the intersection in my path understand the rules associated with traffic lights and are inclined to adhere to them.

6. Arneson (1982:623), for his part, remains fairly agnostic on this question, examining the circumstances in which a "moral obligation to contribute attaches to all beneficiaries" and in which "it is legitimate to employ minimal coercion as needed to secure compliance with this obligation (so long as the cost of coercion does not tip the balance of costs and benefits adversely)." Whether the perceived "costs" and "benefits" are those of the beneficiary or the benefactor, he does not say. Cullity (1995) is less ambiguous on the matter. In his ongoing secondary example of unsolicited (though excludable) "benefits"—wherein a team of "Enterprising Elves," surreptitiously repair shoes left outside one's door overnight (and subsequently send a bill for services rendered)—the "value" of the service appears to be wholly independent of those receiving them in assessing whether these "beneficiaries" are required to pay for it. The work the Enterprising Elves do, he stipulates at one point "is worth the money" (p.10); at another point, he suggests that the recipient of the Elves shoe-repair scheme has indeed benefited since, among other things, "they do not overcharge for their services" (p.14). One can imagine the obvious objection here: for any individual who does not want his shoes repaired, any charge for rendering unsolicited repairs is an overcharge, regardless of the quality of the work, and even—in cases such as this of essentially private, excludable goods—independent of the going market rate for comparable shoe repairs. The libertarian comes to the table armed with this seemingly compelling moral intuition, and it is one that cannot be readily disarmed by

conjecture. Cullity does eventually portray as a general prerequisite for compulsory contribution to benefit-conferring schemes "the requirement that the practice of participation in the scheme is to represent a net benefit *for me*" (p.18, my emphasis), though it's not clear how he draws the general principle from his previous discussion of the Enterprising Elves.

Concluding Remarks to Part I

The headline reads, "'How Bad for the Environment Can Throwing Away One Plastic Bottle Be?' 30 Million People Wonder." Despite the fact that the claim it makes sounds quite plausible, the article it introduces comes not from a major "legitimate" newspaper but rather from *The Onion*, a parody print and web-based weekly published in the U.S. that consistently reaffirms the long-standing literary truth that satire can often tell us more about the world than can our most sober and esteemed official sources of information. In some sense, the headline captures the essence of our environmental malaise from the standpoint of we the comfortable consumers who are no doubt the source of much of that malaise—the existential angst of a people who are, on the one hand, aware that they are collectively harming the planet but who are also, on the other, manically ambivalent about their personal role in that collective harm. One minute we are beating our brows over our undeniably profligate lifestyle and confessing our ongoing collusion in ecological degradation; the next minute we are rationalizing that lifestyle in bits and pieces, pointing out (if only to ourselves) that we and our measly actions couldn't possibly make much of difference, especially when itemized one seemingly "selfish" transaction at a time:

> According to the inner monologue of millions upon millions of citizens, while not necessarily ideal, throwing away one empty bottle probably wouldn't make that much of a difference, and could even be forgiven, considering how long they had been carrying it around with them, the time that could be saved by just tossing it out right here, and the fact that they had bicycled to work once last July . . . All agreed that disposing of what would eventually amount to 50 tons of thermoplastic polymer resin wasn't the end of the world . . . "What's one little bottle in the grand scheme of things, you know?" added each and every single one of them (*The Onion*, 2010).

This is the sort of psychological balm—at once both self-aggrandizing and self-vindicating—that girds the "Tragedy of the Commons" model of environmental politics. We are all equally to blame, yet uniquely exonerated. We are all environmental criminals complicit in the heist, but I was just *one person* standing lookout.

It is, however, a different, equally acerbic headline and associated article from *The Onion* a few years later that I think more aptly evokes how we misunderstand our own participation in environmental harm: "Decision to Circle Parking Lot Produces Carbon Emission That Finally Does It" (*The Onion*, 2014). This one points, with a wry smile, to that hapless person i_k whose environmentally relevant biography I have just related in Part I. He is the improbable but definite individual whose actions tip the scales and render a situation of near-disaster a genuine catastrophe. Were we able to identify him and his oafish act, things might be different. We know deep down that he *must* exist, and that it could in principle be any one of us, but we take comfort in the fact that he cannot readily be singled out and that, therefore, the problems of climate change, atmospheric pollution, water shortages, etc., are bigger than any one of us. It is just a short crawl from this assurance to the one that justifies the single plastic bottle sent to the landfill, and which justifies a coercive moral authority to "save us from ourselves."

I have argued in Part I that this is an unhelpful way to think about environmental problems. The "commons" prone to "tragedy" as the operative metaphor for environmental politics has arguably done a disservice, as crucial as Garret Hardin's essay may have been in pointing out over a generation ago that there must be *something* odd about collective behavior that is at work in environmental problems. As I've suggested in Chapters 1 and 2, the Tragedy allegory has focused our sights strictly on the *jointness* of public environmental goods and thus miscast the moral issues at stake. The Tragedy of the Commons really highlights, I argued, not the conflict between selfish motives and the collective demands of morality but rather the epistemic limits we face in knowing whether we are in fact that person circling the parking lot one more time who, as *The Onion* so bluntly describes it, "finally does it." One cannot conclude from the fact that everyone's contribution has a minimal marginal impact that there is no one who tips the scales; the problem is that it is nearly impossible to know who this might be, and when or where it might occur. Our interest in cooperating is driven as much by a lack of information and pure self-interest in preserving the good as it is in any obligations that might extend to others. Hence the Tragedy alone cannot justify the need for a moral or legal authority to exact cooperation from all; such authority would be redundant in light of what rational individuals would already be inclined to do with full but practically unattainable information about the consequences of their actions.

A better location in which to metaphorically situate environmentalism may be not the pastoral commons but the union shop where, since all workers enjoy the benefits which the ongoing collective bargaining agreement bestows, fairness requires that everyone contribute to its maintenance. This, as was also suggested in Part I, points to the *non-excludability* of environmental goods rather than their jointness; the moral problem posed is not that one can consume such goods without appreciably diminishing the amount available to others (this isn't true, I argued, for *everyone* involved in consuming *crowdable* public goods) but rather that, once provided, these goods cannot be cost-effectively denied to any member of the population that will enjoy them—even those that make no voluntary contribution toward their provision. Hence the rationale for a popular authority in ensuring the provision of environmental goods is not strictly that it is necessary to prohibit selfish actions at the expense of society at large, but rather that it is required to exact cooperation from those who stand to benefit from the provision of environmental goods but who refuse to contribute toward their provision. This would include, in addition to the ostensibly "selfish" free riders who plague the commons, other individuals who claim no desire for the good and who are not driven by selfish motivations. Pleading indifference to such a good, as we've seen in Part I, would be insufficient to absolve one of one's cooperative responsibilities, since those who do make such claims—"honest holdouts"—are effectively indistinguishable from those who aim to enjoy the good but withhold their contribution—true "free riders." Environmentalists have invested a great deal of time and effort in the Tragedy of the Commons metaphor in their attempts to influence environmental policies, invoking the power of the state to limit the "selfishness" and rampant consumerism of Western societies. I think it's clear that, if we are to call upon a coercive authority, it ought to be for purposes of fairness. This requires that we understand the planet not as one big commons but rather as, to adapt the old Industrial Workers of the World (IWW) battle cry, *one big union shop*. Those who play fair and pay their dues are afforded the full benefit of the goods they provide; those who don't play fair and who withhold their dues enjoy the benefits no less, and it is thus permissible to exact a "contribution" from them by coercion. Environmental policies with implications for individual behavior, however complex, need to begin from this simple understanding.

It is also important to point out that there is nothing offered here by way of practical argument about how such collective schemes ought to operate, what would constitute a justifiable form and amount of coercion, by whose authority, etc. The general notion shouldn't need a great deal of explanation, nor should it be couched in euphemism: we are, of course, talking about the role of government and the extent of its proper abilities to tax its constituent citizens and provide public goods on their behalf. How this is done is a separate matter and beyond the scope of the immediate discussion. If one is to claim, however, in the most fundamental sense, that one should be relieved of one's tax burden or exempted from revising one's consumption habits on the grounds that some of the goods provided by these burdens are goods one doesn't desire but cannot easily avoid, I believe we have at least a rudimentary case for claiming that such arguments, however intuitively appealing, are ill-founded.

I indicated in the remarks that opened Part I that the arguments presented here are tentative and conditional. I have by no means demonstrated that any form of the class of arguments I've called "arguments from non-excludability"—be they "simple" or "strong"—are ultimately defensible and resilient to all potential challenges. In fact, I did very little to argue on their behalf at all. Nor did I make any attempt to argue that the "free rider" is himself without moral defense. The gist of Part I is to suggest that, if we think it morally permissible to exact a contribution (by legitimate coercion, if necessary) from the free rider whom we believe desires and *intends* to consume a given environmental good, we may also exact such a contribution from those who claim indifference to the good but will unavoidably be consuming it as well—the honest holdout. I think all but the most strident libertarian agree that the free rider ought to be made to pay; my aim here is largely to suggest that the arguments we think we might have for exempting her counterpart in the honest holdout (namely, arguments from non-excusability and particular forms of arguments from preferences) fail, and that we thus have no morally relevant means for distinguishing them for purposes of environmental policy.[1]

We have also seen that even this conditional argument is further restricted to goods of a particular sort: those which can reasonably be considered important enough—that is of great enough non-subjective *value*—that no rational person could sincerely reject them. No argument is offered here as to how we delineate and value such goods. While I have argued explicitly that purely non-subjective accounts of such value are inadequate for us to decide what is and what is not in the best interest of non-contributors (and that we are thus not justified in demanding contributions from honest holdouts on paternalistic grounds), it should be apparent that any subjective account of value offered by the libertarian is going to be equally inadequate here. To counter the paternalistic appeal to the collective "value" of the good among the cooperators with an appeal to the private "value" of the good to the non-cooperator begs the question no less: the libertarian would essentially be arguing that the honest holdout should not be compelled to pay more than she thinks the good is worth because the good is not worth more than she thinks it is worth to her. Thus while the libertarian is justified in complaining of "paternalism" in purely non-subjective notions of what a good should be worth to the holdout, this doesn't tell us anything about how we should determine the value of the good, and offers no defense for the libertarian "intuition" that it should be entirely subjective.

How, then, should the "economic value" of environmental "goods" be determined? In Part II, we will address this question head on.

NOTES

1. One might quite justifiably find it unsettling that we are resting a call for the intervention of coercive authority on an epistemic limit: it is because *we* cannot distinguish the free rider from the honest holdout that we are justified in treating them the same in the eyes of morality and, ultimately, of the law and of public policy. This is true, but it is arguably not all that unique. Much of what we subscribe to as legal authority amounts to having thrown up our hands, effectively saying, "There is no cost-effective means for us to discern a distinction." This is true not only in areas where elusive preferences and intentions are at issue but also where readily demonstrable *abilities* are at stake. Alcohol retail laws specify a minimum drinking age; speed limits dictate the maximum velocity at which one may propel an automobile; other laws govern the combination of the two. It may be the case that one's twelve-year-old daughter can hold her liquor like a champ and still handle tough curves at high speeds in the family Toyota; we don't have the wherewithal to care, much less to verify this. She will wait until she is the age of her state's designated driving age to apply for a license and drinking age to purchase her Scotch, and she will not mix the two at that point, her unusual tolerances and talents notwithstanding. We are compelled by necessity to lump her into the rest of the population that, on average, exhibits the capacity to drink and to drive only at certain points of maturity, and that can virtually never safely or with prudence do both.

Part II

WHAT ARE WE ASKING WHEN WE ASK "WHAT IS THE NATURAL ENVIRONMENT WORTH?"?

For many people, to stumble upon Bloomberg News, CNBC, Fox Business News, or any similar cable network while flipping television channels during a typical lazy afternoon is like wandering into a meeting of a religious cult. Perhaps it was a similar experience in pre-cable generations to pick up a copy of the *Wall Street Journal's* "business" section (a redundancy if ever there was one) in the waiting room of a doctor or some other professional. Who are these people? Their world seems to revolve around rituals and practices that might strike the uninitiated as bizarre. They speak of this thing called "the market" as if it were some large, unruly circus animal with a mind of its own that is persistently confounding its trainers. World events of grave importance discussed in other news sources are translated into "bottom line" terms, their implication for various portfolios assessed and compared. Things that are valued for their own sake in our ordinary parlance (beauty, trust, esteem, etc.) are monetized commodities in the world of twenty-four-hour business—derived, traded and hedged, their stock's future being the source of speculation for well-groomed people of serious demeanor.

More striking, however, and likely very much akin to the atmosphere of an actual religious cult, is the hubris in the persistent message that *we owe these people our existence*. Those presenters in pressed suits and expensive coifs as well as the captains of industry for whom they speak most certainly would like us to believe that their message is the ultimate truth, that *Wall Street* maps unceremoniously to *Main Street*, and that the various machinations they describe each day are the true gears that keep our pathetic little worlds turning, you're welcome. Naïve romantics may deny it, but everything from wars and famine, natural disaster and heroic acts of charity, down to the petty trials and joys of our daily existence are the mere surface phenomena of the financial core, the equilibrium of buyers and sellers that governs our lives whether we choose to believe it or not. One never feels quite so lectured and summoned to "grow up and face reality" as when one watches the "expert analysis" of the market report at the close of the trading day.

Let's take this milieu at face value for our purposes of assessing the moral weight of ecology. Let's accept as given, for the sake of argument, the psychological claims assumed by the economists who rule over the financial world that people have preferences concerning the environment, that these preferences may be ranked, and that they are more or less "rational" insofar as they are non-contradictory, transitive (i.e., if one prefers A to B and B to C, one prefer A to C), and the like. Let's also assume that individuals might be able to, when prompted, assign to their preferences a common measure such as dollar amounts. Should this matter in how environmental goods are allocated—in whether the adverse effects of climate change are averted, in how much clean air is to be produced, in whether a particular species of bird is to be saved from extinction, in the amount of tropical rainforest acreage that is to be preserved, and so on? Can we move seamlessly from the agreeable claim that different people have different desires and priorities regarding the environment to the claim that the "value" of environmental amenities should be derived and quantified from these desires and priorities? Does, in short, the fact that we might have environmental "preferences" justify the claim that markets ought to coordinate them, and that this process should constitute the foundation of our environmental policies? Economists and other market advocates argue on putatively non-normative grounds that we may answer these questions with a resounding "yes," and many of us may reflexively assent to this under the belief that anchoring environmental policies in what individuals actually desire must somehow accord with our "democratic" predispositions. In Part II, I suggest that this assent would be premature. We will see that the economist is asking us to overlook a good deal of moral reflection that must be done, if only tacitly, before we attach a personal "value" to the natural environment or any particular aspect of it. That reflection involves, in part, many of the sorts of questions we've already seen in Part I—questions revolving around who can permissibly be compelled to share the burden of maintaining a viable natural world and just how that burden should be distributed. Economists and free-market libertarians would like to suggest that these sorts of questions, however valid, come after we've already established for ourselves subjectively the "value" of environmental "goods." I will show in Part II that this view cannot be coherently sustained in the case of public goods such as the integrity of the natural environment. Specifically, I will be arguing that the "willingness-to-pay" measures typically used to mimic private markets and thereby aggregate subjective valuations of public goods, while presumed by their advocates to be "primitive" and pre-moral, are in fact at best a product of our moral reflections regarding the proper distribution of burdens in providing environmental goods. Unlike the value of mundane, private goods that we might purchase on a market, the "value" of the natural environment or any particular environmental "good" it harbors to each of us inextricably involves, in other words, some assessment of how we might fairly distribute the burdens and costs of achieving and maintaining a healthy environment.

Part II is a busy section, and while we will take precise and deliberate steps, there will be several of them. We begin very generally by surveying what is at issue when we believe we are pitting "markets" against "politics."

Chapter 7

Markets versus Politics—a Brief Overview

Imagine that you are approached with a proposition by a neighboring community. The community is home to a lush old-growth forest, with trees bearing potentially useful timber, and its residents have collectively appointed you to be arbiter of how they are to decide to manage the forest. Your job, they make clear, is to decide not the fate of the forest, but rather *how they will decide* to manage it. You privately poll each member of the community and discover that a majority of the residents look favorably upon the notion of "preserving the forest." You also note that the market for hardwood products derived from the forest appears to be thriving. Given this information and nothing more (you are kept ignorant of past management practices, if any), you are charged with the task of determining the most *efficient* form of management of the forest.

Three options might seem salient to you. The first would render the forest *common property* to be shared by the community. All denizens would be free to do what they wish with the trees. Each could harvest as many as he desired, or abstain from harvesting trees entirely, and no one would have the right to restrict the actions of others regarding their use of the trees. Those members of the community who prefer to see the forest untouched—if they wish to see their preferences enacted—would need to persuade potential harvesters of the benefits of a zero-cut policy and foster a cooperative agreement not to harvest, or they would need to strike a mutually agreeable bargain with each person who favors harvesting, effectively "bribing" each to not act on their right to harvest trees. In return for their abstinence in the latter case, each potential harvester (and the number would likely inflate for strategic reasons) would thus require something from each of the non-harvesters—something they value as much as they value harvesting trees.

A second option would be to render the forest *private property*, selling it to the highest bidder (with the proceeds going to the community coffers) who could do with the trees as she sees fit. Given what you know of the community's preferences, this, you reason, is likely to have certain advantages. While the private owner would have the same option to either harvest trees for personal gain or sell the promise not to harvest as in the common-property scenario, she now has the added ability to *exclude* other potential harvesters who were either unwilling or unable to pay for the exclusive rights to the trees. Similarly, the preservationists themselves enjoy a twofold benefit over the common-property situation. One, they need only bargain with one individual (or interest) in order to elicit the promise not to harvest and achieve a state of zero harvesting.[1] Moreover, if zero harvesting was merely a surrogate preference for sustainable preservation of the forest, a restrictive but necessary policy given the conditions for over-exploitation nurtured by common-property ownership, it can be effectively abandoned under conditions of private ownership. Indeed, whereas it would be effective under common-property regimes to *decrease* the demand for wood products, it is useful under privatization to *increase* this demand if the forest is to be preserved and a sustainable harvest is to be maintained. Were zero-harvest advocates to decrease demand for hardwood products under conditions of private ownership, the value of the forest as a forest would decrease. It would, in turn, become increasingly costly for the owner to maintain the land as a forest, as the opportunity costs of not using the land for more profitable purposes—say, a feedlot for livestock—would mount. At some point, it would be more profitable to clear-cut than to maintain trees. It is, therefore, only the sustained demand for wood products that would provide an incentive for the private owner of the forest to maintain it as such. Thus while a common-property regime would compel the non-harvesters to extol the virtues of wood's substitutes and discourage the use of fine hardwood, the private-property regime would allow them to harness the demand for hardwood that already exists.

Yet a third option might present itself to you. The community could submit the forest to *public ownership*, granting full administrative authority to a central governing body elected by each of the community's adult members. The governing body would itself be subject to the strictures of representative government—that is, it must act in the community's "best interests" and, to the extent that is possible to be consistent with this, as an extension of the will of each member. Such provisions would thus grant the governing body full authority to prohibit or permit the harvesting of trees as it sees fit to carry out its duties. It might, for example, fashion a compromise between the competing interests in the community, permitting a level of harvesting that is less than would be under common- or private-property regimes (assuaging the community's conservationists) yet greater than a zero-harvest policy.

Whatever its decisions, however, they would be binding on all members of the community. And while such decisions may reflect the outcomes that might obtain had individuals bargained among themselves, there is no guarantee of this.

How do you go about deciding from among these three options? Both privatization and the establishment of a central governing authority would seem to be the better candidates if preserving the forest indefinitely is the overriding factor in your decision. Either is likely to fare substantially better than the common-property solution, which appears to present many of the incentives you know to be potentially destructive from the "Tragedy of the Commons" scenarios.[2] Thus your job might seem done; you need only turn over an account of your reasoning thus far and a fair coin. Forest-preservation advocates are likely to be disappointed with this outcome, and might rightly point out that if the most "efficient" solution is to be understood as the one which is most likely to maintain the ongoing integrity of the forest, the third option of public ownership might be the singularly best one. Privatization, they might acknowledge, would certainly make it easier than would a common-property regime to render the forest economically attractive and to maintain its commercial value as a source of hardwood products that would prove profitable for the private owner. There are, however, no guarantees that such incentives would preserve the forest indefinitely. Should circumstances change and other uses of the land beneath the forest prove potentially more profitable—suppose, for example, the price of plastic drops sufficiently that it becomes an attractive substitute for hardwood, deflating the demand for the latter—the opportunity costs of not pursuing those land uses (as privatization advocates themselves concede) will climb and the owner will be compelled to clear the forest. This is not likely to happen, forest-preservation advocates would point out, under a public ownership regime. Should the law-making body decide, as an extension of the will of those governed, that the ongoing viability of the forest is in the public interest, a zero-cut or sustainable harvest policy could be made binding and irrevocable, until that time when the majority of the community's residents no longer favor forest preservation. This is, our preservationist would suggest, sufficient reason for endorsing the *public-ownership* model over its private counterpart.

Market advocates would surely counter that it isn't, and contest your interpretation of the choice itself. Your task, they would remind you, is to determine the most efficient form of forest management. The only reasonable way of interpreting "efficient" in this context, they would point out, is that which best preserves the preferences of the community's residents. And it is the play of commercial markets, privatization advocates argue, which provides the better gage for those preferences: "The 'preference' to preserve the forest in the 'public interest' is not a preference in the same sense as a preference for hardwood furniture or fine musical instruments. Only the latter preferences are backed by a willingness on the part of the holder to pay to see the preference fulfilled. The former 'preferences', when backed by the coercive force of government institutions, amount to a distortion of conditions that would obtain had markets been left free to channel consumer preferences; fleeting inclinations that individuals are not otherwise prepared to back with their money or other resources hold sway, instituting outcomes which effectively subsidize the political agendas of environmentalists. It is, in short, only the market for wood products which reveals the true preferences of the community's residents. Professed conservationist sympathies—even if they are in the majority—should not be counted in the same way as the consumer activities of those wishing to use the products the forests provide. If preservationists wish to see their preferences compete in a fair market against the consumer preferences for hardwood, they can determine what value the forest has to them as a standing forest, and be prepared to pay that value on a market."

Many economists would no doubt endorse the concerns of the privatization advocates. Privatizing the forest, they might argue, is Pareto superior to the common-property option, but this cannot be said of the third, public property option. While, in other words, some people are advantaged and no one is disadvantaged by privatization, it is possible that some individuals will be disadvantaged were the forest turned over to the authority of a democratic governing body. This is because, as our privatization advocates have suggested, the "government" solution will be plagued by individuals petitioning the state to achieve outcomes different from those that would obtain under private markets—what the economist refers to as "rent-seeking" activity. Advocates of forest conservation would not need to put up anything they value to see the forest preserved in the same way they would had it been necessary for them to bargain with a private owner toward this end. This, the claim goes, distorts the true markets at work here. Since conservationists are not compelled to pay the full cost of seeing their preferences enacted, trees would be over-allocated in the forest, and the consumer goods made from them would be under-allocated. Such a distortion of the market can be avoided only if the "non-use value" of the forest—the price preservationists would be willing to pay to see the forests preserved—could be established. A private owner of the forest could then appropriately weigh this amount as a cost of forest development: the opportunity costs involved in cutting down trees and harvesting their products instead of striking a bargain with preservationists to abstain from cutting.

The argument for privatization is thus thought to be not an argument on behalf of particular ends but rather on behalf of the means of achieving socially preferred ends—whatever they might be. It acknowledges that something is bound to be lost in a process which translates many individual preferences into collective outcomes, but maintains that markets do a better job of preserving those preferences in the aggregate than do even the most democratic of state institutions.[3] Such a characterization is not, despite the impressions left by contemporary rhetoric, attributable entirely to the policy literature disseminated by the American Enterprise Institute or the libertarian Cato Institute. It is, arguably, a position that has marked a minority strand of classical Liberalism in the last few centuries. The more widely embraced, orthodox account of Liberalism paints an individual divided between consumer and citizen—that is, one who, in certain spheres of life, satisfies (via the market) the desires and preferences which pertain to only him or her, and who, in other spheres, acts upon impersonal concerns as a member of a polity (via politics)—what Marx had described in *On the Jewish Question* as a manifestation of "the secular schism between the political state and civil society." This is the basis of Rousseau's claim that the "General Will" is different from (and superior to) the accumulation of private wills, and the sentiment can be found in Montesquieu as well.

Yet opposed to this orthodox version of Liberalism has existed one that advocates the collapse of this ostensible conflict between self-interest and impersonal interests into the former, such that public life and civil order are no more than the outcome of individuals acting on behalf of their private interests. It is not, to paraphrase Adam Smith's infamous passage, the benevolence of the farmer, the baker and the brewer that provides us with dinner; it is only because the farmer, the baker, and the brewer stand to profit from our transactions. This Smithean tradition,[4] surviving in contemporary welfare and neoclassical economics, does not deny the existence of a "public well-being" so much as it cautions that such well-being cannot be charted *a priori*—hence the claim that Liberalism is neutral in regard to conceptions of "the Good." In contrast to the Aristotelian and Scholastic philosophies it sought to undermine (and which its orthodox counterparts arguably never fully abandoned), this line of Liberalism suggests that "public life" consists in private individuals engaged in voluntary exchanges toward their own ends. Whether we describe those ends as "political" or "economic" makes no difference; they are the individual's preferences to be satisfied, and if there can truly be anything described as a singular social "Good," it must be emergent from (and further facilitate) these private interactions.

What is now loosely referred to as "free-market environmentalism" (I will occasionally shorten this to a less unwieldy "FME") has come to situate itself firmly in this latter strain of classical Liberalism, advocating privatization of environmental goods and market approaches to environmental disputes traditionally conducted in the legislative arena and in the courts.[5] State institutions, free-market environmentalists argue, are an inadequate forum in which to weigh preferences for environmental protection against those for economic development, property rights, and the like:

> The incentive structure in the political sector is complicated because the bottom line depends on the electoral process where votes matter, not efficiency. Because voters are rationally ignorant, because benefits can be concentrated and costs diffused, and because individual voters seldom (and probably never) influence the outcome of elections, there is little reason to expect that elections will link political decisions to efficiency in the same way that private ownership does in the market process. (Anderson and Leal, 1991:16)

Thus the appropriate "solutions" to environmental disputes, FME suggest, resort not to coercive decisions on behalf of the "common good" but rather voluntary transactions carried out by those directly affected by environmental harms and the measures taken to abate them. We are assured that there is nothing un-democratic about turning public policy over to the market; indeed, because it holds individuals accountable to their preferences, it is said to be in many ways a better approximation of democracy than is electoral support for particular policies. A consumer choice, claim advocates of FME, is a "vote" backed by a "willingness to pay"; since such a payment is not required of us when we express "political" interests such as environmental sympathies, the "political" allocation of goods is determined by extra-economic factors and may have little resemblance to the allocation of goods that would result were individuals required to pay for their preferences in the same way they are required to pay for their consumer choices. Despite superficial appearances, the act of voting in a political process is not without cost. The problem with political institutions (as opposed to markets), FME suggests, is that they create the overriding incentive to *shift the burden* of costs. Indeed, in a rhetorical turning of the tables, Anderson and Leal suggest that "political" decisions can impose "externalities" on individuals otherwise prepared to back their preferences individually on a market yet unorganized as a "political interest":

> There is . . . an asymmetry in the analysis of market and political processes because of a failure to recognize that the *political sector operates by externalizing costs*. Consider the reasoning that political agents apply to scientific management. When land is diverted from timber production to wilderness, there is an opportunity cost associated with the reallocation. Private landowners interested in maximizing the value of the resource must take this cost into account in the "price" of wilderness. The bureaucratic manager or politician who does not own the land, however, does not face all the opportunity costs of his decisions. He will take the values forgone into account only if the political process makes him do so. If we assume that the political process worked perfectly (which is the equivalent of *assuming* that markets work perfectly), then the countervailing

> powers of the opposing sides would internalize the benefits and costs for the decision maker.
>
> But there is little reason to believe that the political process works perfectly or even tolerably well. Because politicians and bureaucrats are rewarded for responding to political pressure groups, there is no guarantee that the values of unorganized interests will be taken into account *even if* they constitute a majority of the population . . . To the extent that those who benefit from wildlife preservation do not have to pay the opportunity costs of forgone energy production, they will demand "too much" wildlife habitat. In the absence of a perfect political process, we must depend on good intentions to overpower the special interest incentives built into the imperfect system. This takes a giant leap of faith. (Anderson and Leal, 1991:14–15)[6]

Habit, convenience, and the undue influence of political interest groups ensure that, despite the abilities of markets to efficiently channel private preferences into collective outcomes, political concerns will determine the fate of forests and other environmental issues in most Western democracies.

These remaining chapters do not directly address the question of whether free-market environmentalism is a viable policy program for treating environmental problems and disputes, nor are they about the question of whether privatization of the natural environment is desirable, morally defensible, or even possible. While its conclusions may have a number of implications for those who would advocate privatization as a matter of policy, the primary concern here will be with the normative underpinnings of these policy positions. For FME or the case for privatization in general to be coherent, we would need to understand the environmental value assessments that individuals may offer as *preferences*—somewhat like those elusive mental states we considered in Part I, but used more specifically here by economists to describe the non-normative, private expressions of one's *willingness to pay* for particular conditions in the natural world. Such preferences would have to be like any economic preference we might hold, such that one's "preferences" concerning the curbing of greenhouse gas emissions would be ". . . on the same footing as other preferences—say, the preference that the city of Cleveland continue to support a football team."[7] There is a good deal of latitude and controversy over how such preferences might be assessed and aggregated, and the bulk of Part II takes up the concerns over one of the more common methods of assessing what individuals are "willing to pay" for environmental goods: asking them. We will see that the Contingent Valuation Method (CVM), a survey technique geared toward assessing the value of environmental and other public goods based on one's stated willingness to pay (in yet another acronym, what is often shortened to WTP) for goods in hypothetical "markets," is far from universally accepted among environmental economists, and is thought by some to yield results incompatible with conventional economic value theory (where willingness to pay is based strictly on demonstrated market activity). CVM assesses, these critics claim, not one's willingness to pay for environmental goods as such but rather the willingness to purchase the moral satisfaction of contributing charitably to environmental causes. We will further see that environmentalists and other critics of market approaches in general dismiss CVM as conflating preferences with normative values, forcibly attaching to the latter "prices" that are appropriate and meaningful only for clearly private marketable goods.

While both of these criticisms of CVM raise important normative questions concerning the attempt to "value" the environment in terms of subjective, WTP assessments, I will argue that these are not the normative questions on which we should be focusing. Rather, I think we should be asking whether one can assess what one is "willing to pay" for environmental goods without reference to the amount one believes others too must pay. CVM advocates argue that such concerns are indeed important insofar as the method can elicit amounts that represent not one's true willingness to pay for an environmental good but rather a bid that is intended solely to influence the provision of the good or the amount one would have to pay in its provision. They contend, however, that strategic bidding of this sort can be minimized and rendered ineffectual. I suggest, however, that it is less fundamental to one's WTP bid what one thinks others are *likely to be* willing to pay than what one thinks others *ought to be* willing to pay. And if this is true, it would imply that "willingness to pay" for environmental amenities is not a purely economic, pre-normative preference. Demanding privatization and other environmental policies which treat it as such would thus be misguided.

As with most "signposts" in this text thus far, this last paragraph should strike many readers as opaque and more readily understood after the discussions to come. We move then, from this brief overview of the "markets versus politics" landscape to the matter of just how one might go about asking the question, "What, in monetary terms, is the natural environment worth to you?"

NOTES

1. In his germinal essay "Toward a Theory of Property Rights," Harold Demsetz (1967:357) describes this advantage as "the basic explanation . . . for the preponderance of single rather than multiple owners of property."

2. Note that, despite prominent similarities, there are at least three important differences between this scenario and the classic "Tragedy of the Commons" scenario—each of them intended to be illustrative:

(1) The Tragedy of the Commons typically assumes the "commons" to be the "natural" state of affairs, and points to coordinated outcomes—such as private ownership or government—as rational responses to it (much as did Classical Liberal arguments on behalf of civil society over the "state of nature"). Here, on the other hand, the common-property situation is no less contrived than its counterparts in private markets and centralized authority; the current conditions of the forest are unknown to you, and the "commons" is just another policy choice, to be assessed on the basis of its merits. This allows us to avoid the paradox frequently noted in the Tragedy of the Commons (one that we didn't have the need or occasion to consider in Part I): if individuals on the commons were capable of rational, coordinated outcomes that could transform its property relations and avert overconsumption, there wouldn't be a tendency toward "tragedy" in the first place (see Krier, 1992:337–339, especially the literature cited in fn. 44). The problem is similar to that noted of social contract accounts of civil society: contracts presuppose individuals already inclined to cooperate, and thus can't be considered the beginning of civil society.

(2) While Garrett Hardin's Tragedy of the Commons allegory itself deals with a common pasture, it is frequently employed to illustrate situations where the environmental good in question would involve intractable problems of "ownership"—such as atmospheric, oceanic, or riparian goods which are difficult to disaggregate and parcel out as purchasable chunks. One of the purposes of Part II is to free us from the concern about the strictly logical possibility of privatization. This will allow us to better address particular questions about the normative coherence of the notion of "value" that privatizing the environment presupposes. Thus the good in question here is one for which its scope is more or less readily defined and which can be subdivided in agreeable ways. Granted, there may well be situations in which the two issues are inseparable (a few of which will, for particular purposes, concern us here), and it may be argued that the "existence" value associated with any environmental good is inherently disaggregable. It is sufficient for the general case, however, to emphasize that the question of what gets preserved by privatization and the advantages of doing so are different from the question of whether the environment *can* be privatized, and it may be necessary to assume a "yes" in response to the third question if we are to assess the first two.

(3) The Tragedy of the Commons, as described in Part I, largely assumes rational self-interest to be the cardinal motivator of individuals on the commons. For purposes of practical applicability to a world now populated with a significant number of environmentalists, the scenario here incorporates a definite proportion of rational altruists—individuals whose purposeful activity regarding the environmental good is aimed not at maximizing their own well-being, but rather at achieving particular environmental ends.

3. It is sometimes argued that markets cannot be "democratic" in the same way electoral processes are democratic, since only the latter can claim the egalitarian characteristic of "one-person-one-vote." While markets undeniably allow one to express one's preferences with some effect on society, the argument suggests, one's actual influence varies with what one spends, and this depends on initial holdings of wealth that may be vastly unequal. Nothing I suggest in these chapters will dispute

this argument, though I would point out that it is insufficient in commending electoral processes as uniquely democratic in societies that have both popular elections and the disparities in wealth that can emerge under capitalism. It may be true that each individual in such societies "holds" only one vote regardless of how many dollars she holds, but how many dollars she holds will matter in whether and how she will exercise her franchise (in the historical absence, at least, of strong social democratic parties, as has been true of the United States).

4. "Smithean tradition" may rightly be considered a bit of a misnomer, since Smith himself cannot really be considered a part of it. Both Smith and his close friend, David Hume, maintained the particularity of private economic motivations in one's range of moral proclivities, described earlier in Smith's *Theory of Moral Sentiments.*

5. The argument that markets can or should be used to address environmental problems is not the same thing as the traditional market libertarian argument that markets should be respected in their own right, that any government intervention into free-market exchanges is unacceptable, and that, therefore, whatever the allocation of environmental goods and harms that emerges from market exchanges is the optimal one. While the latter position has characterized the literature of market economists for years (see, e.g., Demsetz, 1967, and Smith, 1981), it is only recently that some economists and policy theorists have come to express concerns for environmental problems and offer market-based solutions as an affirmative, instrumental policy goal—and, hence, term it a form of environmentalism. Terry Anderson, Donald Leal, Roger Meiners, Bruce Yandle, and a host of others affiliated with the Property and Environment Research Center (PERC; formerly Political Economy Research Center) in Bozeman, Montana have been the most active and prolific in defining what they have coined as free-market environmentalism. Though it was Anderson and Leal's 1991 volume which first expressly took up the mantle, the general, comprehensive case was made as early as Hugh H. Macaulay and Bruce Yandle's 1977 study. Meiners and Yandle (1993) followed close behind the Anderson and Leal monograph with a collection of articles—both practical and theoretical—on free-market environmentalism. The *Harvard Journal of Law and Public Policy* sponsored a symposium on "The Role of the Market in Environmental Protection" at the Northwestern School of Law at Lewis and Clark College in 1991, and later published several of the articles presented (most of which are critical of free-market environmentalism) in the Spring 1992 issue (see, among them, Anderson and Leal, Brunet, Krier, Huffman, Blumm, Willey, Stroup and Goodman, Plantico, Funk, and Yandle, all from that year). While not under the label "free-market environmentalism" Steven Edwards (1987) launched an ongoing discussion in the journal *Environmental Ethics* with his defense of market-based environmental policies, evoking responses from Sagoff (1988), Gowdy and Olsen (1994), and Dore (1996). As the usual disclaimer goes, these works barely scratch the surface of what has been written in this area.

6. Mark Sagoff, whose objections to market-oriented environmental policies we will be considering shortly, appears to be unabashedly comfortable with the fact that "politicizing" environmental issues draws into the process the influence of individuals who would seem to have no particular stake in the outcome. "The genius of democracy," Sagoff affirms, ". . . is to let the conflict spread to a larger audience . . . The

genius of cost-benefit analysis is to localize conflict among affected individuals and thereby to prevent it from breaking out into the public realm" (Sagoff, 1988:96–97). Free-market environmentalists would no doubt consider "the feeling of relevance, importance, and community-consciousness flowing from . . . [political] participation" (Sagoff, 1988:96) in issues in which one has no direct interest little more than the psychological consequences of meddling in other people's affairs.

7. This characterization comes from H. Sterling Burnett (1997), a former PERC fellow.

Chapter 8

How Not to Frame the Question

Put aside, for the moment, any conviction you might hold to the effect that one cannot "put a price" on the natural world. If we indulge the economist in the belief that I could indeed determine what the environment is worth to you as a marketable amenity by ascertaining your "willingness to pay" for it, how might we go about this? Certainly, my first step must be to narrow the question somewhat. Rather than aiming at "the natural environment" in general, I would do better to select one "piece" or aspect of it and determine how much you value this. We've already introduced potential bias into the matter, for the ways in which I might divide the natural environment may not be the ways in which you divide it, and the value that you place on the pieces as I understand them may thus not be as reliable as the value you place on the pieces which are meaningful to you. Moreover, in assessing the value that you and, presumably, other living people place on whatever aspect of the environment that we select, we are inevitably skewing the aggregate value toward present considerations. A temporal bias of this sort may be just as troubling, since future generations of people may have something to say about the environmental good in question, and decisions we make now based on the value of the good to you and your contemporaries will be made without assessing their evaluations.

Dwelling on these qualifications will handcuff us to our starting point, so let's put them aside and take a generic, all-purpose environmental good as an example: a lush, old-growth forest much like the one at issue in the last chapter. How might I determine the "value" of such a forest to you? One way in which I might do this is to assess the burdens and costs you endure to enjoy the forest. Assuming that you don't actually live there, I might calculate what it takes for you to pay a visit. Even if the forest is currently provided as a "public good" and administered as a national park (with no or at most a nominal entrance fee), there will be definite costs entailed for you to get there. Distance traveled alone can, all other things being equal, tell me something about the value of the forest to you: I might infer that if you required a $300 plane ticket, a $150 car rental, and another $200 in camping equipment in order to visit the forest, it is worth $650 to you—decidedly more than another forest admirer who needed to spend only $40 for a tank of gas on a day trip. Other things, of course, are not likely to be equal, and the distance traveled (particularly, the costs entailed by distance) alone may leave me with an incomplete picture with which to make absolute or comparative assessments of the forest's worth to you. It may well be that the person taking a $40 day trip by auto to visit the forest is foregoing income from her $750-per-day legal practice, while your week-long vacation translates into only the week's wages at Taco Bell that you'll be missing. The brute costs of moving yourself from one point to the point of the environmental good alone, in other words, would be an insufficient accounting of the good's value to you; I would need to know something about the additional opportunity costs the trip would entail, since you could be doing other productive things during that time.[1] At some point, however, I might indeed be able to aggregate your travel costs, the costs of the motorist's gas, the opportunity costs entailed by both of you, and all of the costs entailed by anyone else who visits the forest in, say, a one-year period, and come up with a latent demand curve ostensibly as useful and as coherent as the demand curve which would result if national parks were actually priced on a market. The economist Harold Hotelling first proposed this *travel cost method* to the U.S. National Park Service shortly after World War II, and it remains in fairly regular use.[2]

As thorough as my accounting may be, however, there will remain some limitations here. Suppose, for example, you've never visited the forest, and have no plans to do so. Would I be justified in concluding from this that you have no regard for it, and that there is no dollar amount which you might place on it? It seems that in focusing on travel costs, I would be assessing the value you place on the forest only if you were among the subset of individuals who were willing and able to visit it, and that what I would be assessing is your willingness to pay to actually be in or make *use* of the forest, not your willingness to pay that the forest simply *exist.* Might not you derive a certain amount of utility, contentment, or whatever we choose to call it, from the fact that the forest is there for others to enjoy or for its own sake, without having directly experienced it first hand and with no overriding desire to do so? This would seem to be possible, and would thus suggest that the travel cost method may underprice goods such as old-growth forests to the extent that it limits its accounting to demonstrated behavior and actual *use values.* Even economists acknowledge that one may be willing to pay for the *existence* of particular environmental goods that one might have no intention of directly experiencing. Such non-use values may be attributable to the desire to ensure possible future use (what economists call *option* value), to vicarious consumption (i.e., satisfaction in the knowledge that others may consume the good, either current or future generations—what economists refer to as *bequest* value) or to stewardship (i.e., satisfaction in the existence of environmental goods for their own sake and regardless of whether anyone consumes them—what is described as *existence* value).[3] Either way, I may not capture such values if my method is limited entirely to waiting for you to get up and travel to the environmental good with the hopes that I might tally your sales receipts along the route.

One way in which I might avoid these limitations the travel cost method might impose upon my ability to assess how much you value the forest would be to ask you straightforwardly "What is it worth to you?" While this method has the decided advantage of simplicity, such simplicity might well belie a number of complications. How, for example, would I word the question? To simply inquire "What is the forest worth in dollars?" is sure to invite problems. You might, for example, protest that the value of the forest (or forests in general, or the natural environment in general, for that matter) is immeasurable and can't be pinned down to a dollar amount. I could play advocate for the economist here, and suggest that often, when we don't know or can't articulate the value we assign to something, we compare it to things we think we *do* know the value of. Typically, these things are commodities—items for which we have a rough idea of how much one unit of them will fetch in *other* commodities which we value. In lieu of the enormous task of accounting for such hypothetical exchanges, I explain to you, we translate this relative exchange value into a common denominator: money. This could, I suggest, despite our convictions and noblest environmental intentions, be applied to environmental goods as well. You might be persuaded by such an argument, yet whatever "value" you may quote here is not likely to be reliable under such conditions given the level of abstraction the question involves.[4]

Somewhat more tangible than "What is the forest worth?" might be the question of what you would be *willing to pay* for the forest. While this question would better conform to standard measures of value in welfare economics, it's still a bit abstract to work with. You might, for example, understand the question as one of what you'd be willing to pay to *visit* the forest, and I'd be no better off than if I were relying on the travel cost method. More directly, I would probably want to ask you what you would be willing to pay to ensure a certain state of the forest—regardless of any intentions to experience it first hand, and given that there are competing options for which others might be willing to pay. Why is this so? Willingness-to-pay measures in welfare economics generally assess not, as we might imagine, what individuals are inclined to spend in order to reach "higher states" of happiness, satisfaction, etc., but rather what they are inclined to spend to maintain a state of *indifference*—an equilibrium state in which we have only as much of preferred goods as we desire, no more, no less.[5] For me to assess the value to you of any of those goods which you currently enjoy, I would need to introduce a potential or hypothetical loss in the amount of one of those goods, or effectively "remind" you that you currently don't have as much of the good as would make you content. Hence in this case I might speculate that the forest could be given over to clear-cutting so that the land on which it sits could be put to more profitable uses, and ask you what you'd be willing to pay to prevent this and to see the forest maintained as such. This is essentially what constitutes the survey method most commonly used to assess the subjective value of environmental goods: the *Contingent Valuation Method*. The solicited values are *contingent* upon the particular, hypothetical conditions specified in the question—conditions which stipulate some change in the availability of an environmental good and solicit the amount the respondent would be willing to pay to ensure or avert that change.

So, my objective is a bit clearer to me now. I need to stipulate a definite albeit hypothetical set of conditions concerning the availability of the forest and the possibility of clear-cutting it, and ask you what dollar amount you would be willing to pay to ensure its ongoing viability. If you're a friend of the environment and a fan of the forest, this amount would presumably be substantial; if your ongoing peace with the world doesn't really require that you share it with this forest, the amount would presumably be negligible, if it is indeed non-zero. I aggregate your willingness-to-pay with that of others and I derive an amount which might be said to represent the economic "value" of the forest. Is it all that simple? Given what we know about economic preferences and about survey research, not likely. A number of ambiguities are likely to threaten the stability of the willingness-to-pay estimates I solicit through CVM. Unless I'm careful in designing my survey, I may not have valid and reliable data (and even if I were to exercise caution, I might still have doubts).

For one thing, the situation is likely to be a fairly novel one for you. Even if you've made major purchases before, and even if some of those have involved the purchase of property (recall that I'm not necessarily asking what you would pay to *own* the forest), you've likely never been accosted with a question of this sort before—certainly not in a format where you are provided with limited information and asked to make a bid on the spot. Estimating what we would pay for something typically involves a quick decision over a minor consumer commitment (pasta at $x per pound or pasta at $y per pound) or a long and thorough decision process for larger commitments. Asking you what you would be willing to pay to ensure the viability of a forest—let alone less-tangible environmental goods such as averting the adverse effects of climate change—is likely to tread on unfamiliar consumer territory for you, and this might well make whatever estimates you offer somewhat suspect.[6]

There are a number of things I might do to provide a little more context and make the question an easier one to grasp and which might, in turn, render your response a more reliable marker of the information I'm after. I might, for example, make the question somewhat more concrete by providing substantive choices in just what it is you'd be paying for. Rather than inquiring about "the forest," I might prompt you with accounts of different states of the forest, varying in terms the degree of clear-cutting and the corresponding proportion of old-growth trees. Hence the more abstract question could be pinned down to one of what you are willing to pay to move from a relatively lower proportion of old-growth content to a higher one.[7] I might even stipulate the particular amenities you'd be paying for in the forest—both use and non-use—and lay out a variety of scenarios with varying levels of each. A state of the forest which allowed x amount of recreational use, y amount of wildlife preservation, and z amount of atmospheric exchange might thus have a different value to you than one in which these factors combined at levels other than x, y, and z.[8] Prompting you with options of this sort may well help you articulate just what the forest is worth to you.

Similarly, an open-ended question about the amount to be paid might not be as readily answerable as one which proposes a value and asks if you are willing to pay it. "How much would you pay to ensure the existence of the forest?," in other words, might see a different range of responses compared to "Would you be willing to pay $x to ensure the existence of the forest?" A "no" response to the latter need not close the bidding; it could prompt me to lower the proposed amount incrementally until I reach $0 or a "yes" response. While willingness-to-pay amounts in the close-ended, discrete-choice situation might be biased by an "anchoring effect" (i.e., a tendency for responses to cluster around the amount I chose as the opening "bid"), empirical evidence in CVM surveys suggests that open-ended responses tend to be systematically larger and somewhat less reliable than their close-ended counterparts.[9]

Whether I leave the question open-ended or prompt you with an amount and ask about your willingness to pay it, what is of interest to me is not just *any* amount that you are willing to pay but *the maximum amount*. That you might be willing to surrender a week's wages for the forest tells me something, but unless I know that you wouldn't go any higher, it's not quite telling me the *value* of the forest to you since, for all I know, you *might have been willing to pay more*. The same is true if I instead asked you what it would be worth to you to *not* enjoy an environmental good. Suppose, for example, that rather than asking what you would pay to see the forest avoid the saw blade, I ask you what you would accept as compensation to allow the forest to be cleared for other uses. This should, we would think, be the mirror image of the positive question: the maximum you would be willing to pay for a good should be symmetrical and effectively equivalent to the minimum you would be willing to accept to forego it—at least insofar as the welfare economic theory that underpins these fully subjective notions of "value" would predict. If the highest amount we would be willing to pay for a good indicates its value to us, the lowest amount we would be willing to accept as compensation for not having the good (i.e., our "willingness to accept" or WTA) should be a similar, if not equivalent indicator of that value.[10]

Oddly enough, the anticipated symmetry between willingness-to-pay and willingness-to-accept measures is consistently absent in CVM research. The accumulated empirical evidence suggests that individuals tend to place a higher value on an environmental good when they calculate the minimum compensation they require for being "deprived" of it than when they assess the maximum they would pay to secure it (indeed, protest responses of the sort "there is *no* amount . . ." are more likely to be elicited in reference to compensation than payment). It's not clear, moreover, whether this empirical artifact represents an inconsistency in the CVM method or whether it is an accurate reading of an anomalous economic phenomenon.[11] Either way, it's not likely I'm going to get a reliability check of the value of the forest to you by asking you both the maximum you'd be willing to pay to preserve it *and* the minimum you'd be willing to accept to forego it.

Even if I could arrest each of these potential sources of imprecision—and the empirical evidence on this is anything but promising[12]—there remains one area of ambiguity arguably more fundamental to the question at issue here. "Just what," you might well ask, "are you proposing that I pay? You've narrowed the question to a particular state of the forest, and even provided a series of potential 'bids' of which I might approve or disapprove. What would my payment of an agreeable 'bid' amount to, though? Am I purchasing this myself, as I might purchase a jar of peanut butter? Am I quoting an amount I might be willing to pay provided everyone else agrees to contribute this or some similar amount? Or am I to understand it as a charitable contribution which would ease my conscience without necessarily having any practical effect on the state of the forest?"

Your sustained confusion here would be understandable, for I've included nothing in the question thus far that would indicate what your payment, regardless of the amount, would represent. Any dollar amount you might quote or assent to is bound to have a different meaning in terms of the forest's "value" to you depending on the vehicle of payment you understood the question to entail, and the same dollar amount quoted by different individuals can represent entirely different accounts of "value." Suppose, for example, you understood the question as referring to the "market value" of the forest, and took "willingness to pay" to describe the *purchase* of exclusive rights to the forest. You might assess this value at, say, $100 dollars. You realize, of course, that the question is a hypothetical one, and that you are not making an actual bid; you are merely proposing what you *would* pay for ownership of the forest (or, at least, sufficient control over its fate) if it could be auctioned off on a market basis. You realize, moreover, that there are others who would likely bid more for the forest on this hypothetical market and that, given other market prices, it would probably require more than $100 to make any substantial effort at improving or even maintaining the conditions of the forest. Nonetheless, $100 is the maximum that you would pay for the exclusive right to ensure that the conditions of the forest described in the question obtain.[13]

Suppose your friend Barney, in response to the same question from me, also puts the value of the forest to him at $100 dollars. Barney, however, understands the question not as one of individually purchasing the rights to the forest on a market, but rather as a tax referendum. The $100 represents for him the additional amount that he would be willing to pay in his local tax bill every year in order to prevent the forest from being clear-cut.[14] More important, however, Barney is assuming that others would be paying the same increase in their taxes during the same period. Unlike the case of a private market purchase, Barney's assessment of the forest's value here is based upon not only what he would be willing to pay for the stipulated conditions to obtain, but also upon what he believes everyone else should pay. That others may not "value" the forest in the same way he does and thus may not be willing to pay a $100 tax hike to ensure its viability is another matter, as is the fact that Barney may not have a definite scope in mind as to how many individuals the tax increase would affect (if Barney's assessment of the forest is based primarily on its option or bequest value, such a scope might be fairly local; if, however, it is based on existence value, he could coherently believe everyone on the planet ought to be paying for the ongoing care of the forest). Whatever that scope is, however, it is a scope greater than that of Barney alone, and his assessment of $100 cannot, like you, represent the total "value" of the forest to him but rather the portion of the costs of achieving the stipulated conditions that he is willing to bear.[15]

Suppose further that, confronted with the same question as presented to you and to Barney, your friend Moe expresses a similar willingness to pay $100 on behalf of the forest. Moe, however, understands the $100 to represent neither the market value of the forest to him nor a fair tax burden that would be shared by others at the same rate, but rather a charitable contribution. Somewhat like you, and unlike Barney, Moe's assessment of $100 is made without regard to what he believes others should pay to preserve the forest (though neither you nor Moe, as will be discussed later, would likely make such an assessment non-strategically—that is independent of what you believe others like Barney *are likely to* contribute). The $100 is simply what he would pay to achieve the "warm glow" of satisfaction in knowing that he contributed something toward a goal he thought to be a noble one. He may realize, of course, that this goal might well bring him no returns beyond that "glow" if few others make such a "charitable contribution," but lacking here is Barney's notion that others *ought* to be paying the same amount he pays (i.e., he sees that act as supererogatory rather than obligatory). At the same time—much like Barney and unlike you—Moe doesn't see his offer as capturing what he thinks the "total value" of the forest to be, just what he's willing to pay "for his part." Unlike the situation where you understand the question to be asking about the market price you would be willing to pay for the forest (and, presumably, not clear-cut it), Moe does not understand the question to be one of assessing the forest's value *per se*, merely what he'd be willing to *contribute to the cause* of seeing it in a particular state.

In practice, of course, CVM surveys do not leave the payment vehicle in this state of ambiguity. Respondents are typically provided a detailed description of the environmental good in question, the change to which it will be potentially subject, and the means by which their hypothetical (or, in the case of some experiments, actual) "bid" to affect or avert that change would be collected. This presents at least a superficial problem, however, for free-market environmentalists hoping to hinge the allocation of environmental benefits on voluntary market exchanges. Recall that the essential argument from FME advocates is that privatizing environmental goods more efficiently manages them (in terms of translating the preferences of many individuals into public policy) than does a political process, and that this is sufficient reason for doing so. The private allocation of such environmental benefits, in turn, depends on assessing their market "value," and this requires that we aggregate individuals' WTP for these goods. Yet, as we have seen, there is no impartial means of determining this WTP—that is, no method which does not require that we lay out specifically whether private market functions would be an effective vehicle in carrying what one is willing to pay or if political means such as taxation would be the only ones feasible. The means of allocation, in other words, is something the economist must bring to the CVM situation, not something we can derive from it. Thus apart from the question of whether aggregate subjective WTP is an adequate assessment of the "value" of environmental goods, it should be clear that there is nothing inherent in this WTP which would suggest that the private market allocation of such goods is what individuals have in mind when they are probed by economists. The question "What amount are you willing to pay for x?" especially where x is an environmental benefit, must convey much additional context for it to make sense, and part of this context must be an account of whether one would be making a market (or market-like) purchase, whether one is describing what one believes to be a fair tax increase or reallocation (or, to put this in the language we used in Part I, whether honest holdouts and thus, inevitably, free riders would be tolerated), or whether one is making hypothetical contribution to a charitable cause. Nothing in the fact that individuals might be willing to make such a WTP assessment, or in the fact that we can solicit one, implies that environmental goods are best or even easily allocated as private goods on a market.

We have thus potentially disarmed at least one critical portion of the essential FME argument. If it is not possible to solicit market "bids" for environmental goods without also laying out the conditions of such a market, there is nothing inherently enlightening about such bids as to the "true" value of the environment. The economist and privatization advocate tell us that the only thing separating clean air and cans of soup in terms of markets are conventions, taboos, and wrong-headed policies that stamp the former as "unmarketable" and the latter as "marketable." Yet, in soliciting WTP "bids" for environmental goods, we are compelled to lay out precise circumstances of how such bids would be collected, whether others would be paying the same amount by mandate, whether one's bid will afford one exclusive rights to a good or simply ensure its availability (or even its mere "existence") to everyone, etc.—things which don't seem to be required in soliciting WTP assessments for cans of soup, athletic shoes, and the like. We are, in other words, invited to believe that environmental amenities are no different from other goods we can allocate in a capitalist society, and we are told that such a belief is justified by the fact that people can and will offer more or less reliable bids when asked what environmental goods are worth to them. But for such bids to have any meaning, we must tell a fairly elaborate story about what will become of them. To accept the FME case, we would have to accept the claim the apparent differences between typical consumer goods allocated on a market and public goods are really no more than apparent. Yet environmental "goods," like all public goods, while eluding hard and fast categorization (as we saw in Part I), do tend to cluster at the extreme ends of the jointness and non-excludability criteria, and this must surely have some impact on the way we assess the "value" of those goods to us. Nothing in the way CVM surveys must be administered suggests otherwise, and any argument to the effect that we can assume the marketability of a good by virtue of our ability to solicit WTP assessments alone will thus be inadequate.

This in itself, however, would not constitute grounds for dismissing the role of CVM entirely, let alone for abandoning the notion that the value of the natural environment in social policy should somehow be based on individual, subjective value assessments of environmental goods. In the remaining chapters of Part II, we will consider the concerns of economists and environmentalists alike that CVM cannot reveal the content of WTP assessments for environmental goods; we can consider these the "internal" criticisms of

CVM by those who retain the belief that markets can somehow register the true value of the environment. We will see that while such concerns are often misplaced, they are instructive in shaping an "external" critique of CVM and indeed of the very notion that a market "price" can meaningfully be placed on the natural environment. There are, as we will discover, good reasons for believing subjective WTP bids are not adequate elements for assessing the value of the natural environment in public policy.

NOTES

1. For a more complete account of these costs, see Johansson (1987:119).

2. See Hanemann (1995:80–84) and Adamowicz (1995:145–146) for background and an account of the extensive literature on the travel cost method, as well as Gardner Brown Jr. (1993) for a practical comparison of the travel cost method and the Contingent Valuation Method, which we will be considering shortly. The travel cost method of establishing the "willingness to pay" of consumers for environmental goods is one of a number of methods that are typically subsumed under the heading *methods of revealed preferences*. These generally assess WTP based on behavioral criteria and demonstrated (i.e., revealed) tendencies to actually expend known costs in exchange for particular goods. Another method of "revealed preferences," *hedonic pricing*, is also commonly used to assess the value of environmental goods, although it tends to preclude the types of goods which lend themselves to the travel cost method and focuses on more enduring amenities such as air quality in a particular region. Hedonic pricing, moreover, relies more heavily on statistical abstraction than do methods like travel cost assessment. We might, for example, note that individuals in areas with higher air quality tend to pay higher property taxes. In order to attribute this willingness to pay higher taxes to the amenity of better air quality, we would need to hold constant (typically via statistical methods such as multiple regression analysis) as many other factors as possible which might contribute to higher property taxes, such as better employment opportunities, regional climate, proximity to other natural attractions, etc. Thus while the travel cost method might be described as "indirect" insofar as it takes as an indicator of WTP something that the agent herself might not even consciously register as declaration of her preferences, hedonic pricing methods may seem even somewhat more speculative, insofar as willingness to pay for the good in question must be teased out from among other competing hypotheses for why a given amount is actually paid (to put this differently: as imprecise as the travel cost method might seem, we aren't likely to attribute a vacationer's expense of $1000 to visit Yellowstone National Park to the quality of the food concessions alone; we can, at least, make some assertion about his willingness to pay for the natural amenities of the Park itself). See each of the articles cited above as well as Johansson (1987:93–114) for more on the hedonic pricing method.

3. Mitchell and Carson (1989:59–67) provide a fairly comprehensive summary of the literature on how the distinction between *use* and *non-use* values such as *existence*

values are drawn, including the extensive debates among economists over whether existence values based on the perceived "intrinsic" value of nature should be admitted or rejected as "counterpreferential." Much of this discussion is at the heart of the normative issues to which we will return later in Part II. It should also be noted here that, despite the similarity in terms and minor overlap in meaning, the distinction between *use* and *non-use* values as it is understood in this context is not the distinction between *use* and *exchange* values in classical and Marxists economics.

4. Undervaluation of the good would be a pronounced tendency, but it would not be the only tendency. One might surmise that you would be just as likely to "pad" your bid in order to "play along" with the perceived expectations of the question. Survey researchers have long noted the tendency of interviewees to offer what they believe to be "socially desirable" responses to survey questions, often independent of their true inclinations.

5. This is essentially the same sense of "indifference" we were using in Part I to describe the true (but difficult to discern) dispositions of the honest holdout; she was *indifferent* to the provision of the environmental good and to the cooperative scheme that could provide it insofar as she was at a point where she preferred to expend nothing on it. The cooperator, on the other hand, would be willing to expend something of value to reach that same state of *indifference*—as might the free rider if he genuinely believed he could not get away with withholding his contribution or that the good would not be provided without his contribution.

6. See Mitchell and Carson (1989:35) for a summary of some of the literature which suggests experience via repeated exposure to CVM survey tools in experimental situations might render estimates more reliable.

7. The question could not, of course, be entirely literal in reference to a single forest in this example: though we might control the rate of harvesting, we can't add older trees to an already-existing forest. This method is, however, common when amenities such as improved air or water quality are at stake. Researchers typically display landscape photographs depicting varying degrees of pollution and solicit the respondents' WTP to move from one state to the next. See Johansson (1987:Chapter 7) for an example of this technique.

8. *Stated Preference* methods of this type, adapted from *Conjoint* and similar methods used in market research, involve a list of attributes associated with an environmental amenity and their absence or presence in different scenarios. Respondents are presented with a subset of various combinations of the levels of each attribute (there are statistical methods of eliminating the need to present all possible combinations) and asked to rank these states in order of their willingness to pay to obtain them or, effectively, "purchase" the environmental good with this particular combination of attributes. See Adamowicz (1995) for an example of this method.

9. See Mitchell and Carson (1989:97–104) on these variations in CVM survey designs.

10. The reader unfamiliar with the economist's logic might rightly ask here, "Why should anyone be asked—much less *willing*—to accept anything for the imposed *loss* of an environmental good that already exists and is doing fine on its own? Isn't this essentially what we mean by 'pollution' or 'environmental harm'? If

someone is imposing such a denial and depriving me of something I already have, don't we consider this *illegitimate* and restrain the perpetrator by passing laws that prevent him or her from threatening the environment further? Why is money an issue here at all?" While this is not a matter I will be pursuing in any depth in this book, it's a powerful and perfectly reasonable intuition, and addressing it has in fact been a prominent effort in legal and environmental discourse for nearly a century. The principle economic alternative to simply "outlawing" pollution was first outlined in the 1920s by Alfred Pigou (1950) in his *Economics of Welfare*. Pigou had described pollution as a market failure—a *negative externality* of the sort outlined here in Chapter 1—that is imposed upon society in the course of otherwise-productive activities. In lieu of an attempt to enjoin the polluter to stop his pollution, however, Pigou advocated a system of taxes to be levied against producers of pollution which would effectively *internalize* the externalized costs. Externalities such as pollution, according to Pigou, are an attractive option only because they are priced under their market value; since the polluter himself doesn't pay the full cost of its effects, pollution tends to be over-allocated. Pigouvian taxes—what are now often called "green taxes" by contemporary advocates of Pigou's approach (see, most notably, Hawken, 1993)—are thus not an attempt to usurp the market on behalf of a clean environment but rather a means of ensuring that pollution and other externalities are produced at no more than a true market would allow had it the ability to operate without externalities.

The "Law and Economics" school represents the most prominent economic challenge to the notion that negative externalities require the intervention of the state—even in the indirect form of enacting "green taxes" on polluters. The linchpin of the Law and Economics school is what has come to be known as the "Coase Theorem," developed in Ronald Coase's now famous (1960) article, "The Problem of Social Costs." Stated about as succinctly as one can state it without loss of meaning, the Coase Theorem simply suggests that the legal question of whether the polluter is within her rights as a producer or whether the "pollutee" has a right to enjoin the polluter's behavior is superfluous: there is a socially optimal amount of pollution that the former may impose upon the latter regardless of where the law stands on the matter, and if transaction costs can be kept below that point, parties to the affair have a mutual interest in bargaining to locate it. Coase assumes, without serious objection, that the polluting firm's marginal profits decrease as units of pollution increase; hence rather than maximizing production and the pollution that comes with it, it is optimal for it to either restrain production and accept compensation from the "pollutee" for lost production (if indeed the law favors the firm) or to compensate the "pollutee" to the point where it is still profitable at the margin to produce and pollute (if the law favors the "pollutee"). To put the Coase Theorem in Pigouvian terms, if actions you take with your property impose negative externalities upon my property (thus devaluing it), it is in my interest to seek compensation or to persuade you to cease such actions with financial incentives by *compensating you* for foregone production; where we settle in this range will be the same regardless of which of us is in the "right" in the eyes of the law. The amount I might "owe" you, if this were indeed how we settled, would be a direct reflection of the amount your property is devalued by ceasing the productive yet cost-externalizing activity; it is, hence, compensation for the losses you incur by cooperating with me. To force you to cease your activity without such compensation would, according to the Coase Theorem, constitute takings—the unwarranted confiscation of your property (or, at least, its value). Environmental examples, of course, lend themselves to easy manipulation by the Coase Theorem, and have provided the classic articulation of its practical benefits. If you have invested a substantial portion of your resources in a particular productive activity only to be told that a new piece of environmental legislation would prohibit it due to its downstream effects on my property, few individuals might dispute that this would seem an undue burden upon you—a taking of your property. It would be better for each of us, it would seem, for me to either assess the lowest value of my property at which I'd be willing to sell it to you (in the hopes that it approximates the highest amount you'd be willing to pay) or to compensate you for lost profits you incur by ceasing the activity. Since any normative scale on which we might rank our respective activities is irrelevant according to the Coase Theorem, neither activity, productive or otherwise, is any more "valuable" than the other. If your activity produces side-effects which impede mine in some way, it would be more profitable for each of us to consider which of our activities should be continued and which of us should cash out at that point. This, Coase claims, is the insight that renders obsolete conventional legal approaches to environmental spillovers. It has bewitched libertarians for decades, and even those not particularly inclined toward libertarian ideals might be drawn to it—at least until one raises moral concerns about the logical implications of it all: are laws against larceny, after all, thus merely an unwarranted overreach by the state that prevents thieves from bargaining directly with me over how much I might compensate them for not stealing laptop computer?

The "Law and Economics" literature is now voluminous, with its particular approach to legal matters dominating many of the articles published in the *Journal of Law and Economics* (though the journal itself arguably predated the field it for which it now serves as an eponymous forum). For comments on and fairly comprehensive elaborations of the Pigouvian position and the implications of Coase's writings for it, see Turvey (1963), Davis and Whinston (1965), Mishan (1965 and 1971), and Knapp (1970); Hahnel (2011:54–9 and 106–12) provides an especially lucid explication, with helpful diagrams, in the contemporary environmental context. Hanly (1992) raises a number of relevant ethical objections to the Coase Theorem, and Epstein (1985) explores the property dimensions of the debate. Sagoff explains eloquently how the problems of the Coasian approach when applied to environmental valuation are rooted in the very "hypothetical" nature of markets for environmental goods: "the big difference [is] between traditional markets, in which property rights are well defined, and 'contingent' markets, in which resources go, in principle, to the highest bidder" (Sagoff, 1988a:59); in traditional markets, participants are allowed to refuse to sell their property to other participants willing to pay the highest price.

11. Mitchell and Carson (1989:30–38) provide a review of the extensive debate that has emerged around this issue, at least as it was conducted until the publication of their volume in the late 1980s (see also Knetsch, 1990), and provide a summary of

the various explanations of the WTP/WTA asymmetry that have been proposed. These often revolve around potential psychological differences between "having" something that is somehow lost and acquiring something we do not yet have (although the state of "indifference" toward which welfare economists suggest we strive would entail that there should be no psychological difference here). We might, for example, suppose that relinquishing an environmental good that we now "hold" or to which we are "entitled" will have greater value to us than would the same good if we were merely considering "purchasing" it. The economist E. J. Mishan (1982:63–64) provides a characteristically lucid example of a situation well outside the realm of environmental goods in which we might also anticipate an asymmetry between WTP and WTA measures. He invites us to imagine a man burdened with imminent death from a fatal heart disease and presented with the option of an expensive life-saving surgical procedure. Such a man, he suggests, would likely "value" his life (in terms of his WTP) at whatever amount of money he could get his hands on to pay for the surgery—a potentially large but finite amount. Should, on the other hand, a member of the RPAAC (Reduce the Population at All Costs) society approach him with the question of what he'd be willing to accept to forgo the surgery and await his death, it would be difficult for us to imagine an amount upon which they might settle.

12. Diamond and Hausman (1993), for example, conclude that the accumulated limits of survey methods render CVM generally unreliable in assessing WTP for environmental goods; Hanemann (1995) disagrees with this conclusion, and suggests that there are techniques available to make CVM surveys sufficiently reliable. Mitchell and Carson (1989:Chapter 13), in their own evaluation of CVM, can be situated in the latter camp, as can Kenneth Arrow, et al.(1993), whose National Oceanic and Atmospheric Administration panel report generally affirmed the usability of CVM surveys under particular circumstances. It is important to distinguish, however, the "external" criticisms which suggest that we cannot put an economic "value" on the natural world from the "internal" criticism that the data CVM collects cannot comport to the assumptions of subjective value proposed by orthodox economic theory. We'll see later in Part II that each critique has distinct normative concerns with CVM.

13. There are situations where you might be given the opportunity to "bid" on a market transaction concerning the forest yet not estimate the value of the forest in terms of what you would be "willing to pay" for it *simpliciter.* Suppose, for example, the threat of clear-cutting the forest comes from a power company that wishes to purchase the land and extract the abundant deposits of coal beneath it at a profit sufficient to lower your energy bill. The question might become one of what you would be willing to pay in increased energy rates—more or less market rates, if we set aside the irreducible political content of these in the real world—in order to see the forest preserved. Hence the WTP value serves as a proxy for the market value to you of the forest, but this would be very different from the situation in which you understood the question to be "what are you willing to pay for the forest?"

14. Even this might involve potential ambiguity: does stating a willingness to pay $100 dollars per year in taxes to affect particular conditions in the forest amount to permitting an *addition* to one's tax bill or an expression of one's desire for the *reallocation* of tax budgets to accommodate the money spent on the forest? If the

latter, is the stated preference implying that the money be taken uniformly from all other areas of tax expenditure, or are some areas more vulnerable than others?

15. Barney's understanding of the question has some support in the empirical literature on CVM. The Arrow et al. (1993) report concluded that voting (tax referenda) scenarios with close-ended questions were the most reliable of CVM formats.

Chapter 9

The Embedding Effect and the Moral Content of Economic Valuation

Is saving a wilderness area of 10,000 acres more valuable than saving one of 5,000 acres? If so, is it worth twice as much? Similarly, is reducing the content of a particular airborne toxin from 24 parts per million to 6 worth three times as much as reducing it to 18 parts per million? What about our recurring forest example? Would preserving an old-growth forest with 200-year-old trees be half as valuable to us as preserving one which has trees averaging 400 years in age?

Judging from the volume of literature devoted to what is labeled the *embedding effect* in CVM studies, one might conclude that environmental economists anticipate linear extrapolations of this sort from an individual's WTP assessments. The embedding effect is an empirical artifact of CVM surveys in which these expectations are thwarted—in other words, in which there is an observable tendency for "bids" on more inclusive environmental goods to be only slightly larger than less inclusive ones, if larger at all. Embedding may be simply quantitative (e.g., the WTP to preserve a single timber wolf being about the same as the WTP to preserve the entire species), categorical (e.g., the WTP to preserve the timber wolf being about the same as the WTP to preserve all endangered types of wolf), or temporal (e.g., the WTP for five-year plan to restore a timber wolf habitat to its "natural" state being about the same as the WTP for a ten-year plan).[1] In any case, the implication is that the marginal value of the amenities offered in the more inclusive good or bundle of goods—goods which we might otherwise presume to have some subjective value if assessed on their own—is roughly zero. The phenomenon is widely observed in CVM studies, though it is generally considered anomalous to the expectations of the welfare-economic theory on which CVM is based. This has made it a problem to be explained.

One might well wonder why economists or anyone else would expect a linear correspondence between "units" of an environmental good (including those cases where "units" amount to abstractions) and an individual's WTP or, to put it differently, why marginal "units" of an environmental good would be uniformly "priced" at all levels of its availability. Environmental goods, as even economists concede and as I've described in Part I, tend to defy allocation in the way that commodities are typically allocated in capitalist societies. It is conceivable that someone willing to pay x for a can of soup would be willing to pay 5x for five cans of soup. It doesn't seem as obvious, however, especially given the concerns raised in the previous chapter, that someone with a stated willingness to pay x to reduce greenhouse gases to their 1990 levels in 10 years would pay 2x to reduce greenhouses gases to the same level in five years. To what do we attribute this apparent difference? We might be tempted to chalk it up to easy metrics: consumer goods allocated on a private market can be disaggregated and parceled out in neat, measurable units, while environmental goods are "larger," effectively disaggregable, and tend to approximate the conditions of *public goods*. As we've seen in Part I, however, we probably don't want to hinge this difference on something "inherent" in the goods themselves. What makes processed soup packageable and saleable in cans is the fact that someone can do so at a cost less than that which individuals are willing to pay to secure exclusive rights to some unit of processed soup (which is, as odd as it sounds, essentially what we are doing when we purchase a can from the local supermarket). What makes the benefits of reduced greenhouse gas emissions less amenable to such packaging is likely that the cost of allocating it in units to which individuals would have exclusive claims (in the language we used in Part I, from which they could *exclude* others) is prone to be prohibitively higher than anyone would be willing to pay for the private allocation of such a benefit. Hence the inexperience most of us have in assessing just what price we might pay for such a "package" would suggest that neat, linear demand functions are going to be hard to come by for most individuals when prompted for WTP amounts.

This line of reasoning suggests that the embedding effect may not be surprising given the elusive nature of allocation involved with public goods such as environmental goods. One might further suggest that embedding is not only unsurprising but consistent with what we might expect from *any* type of good. We know, for example, that demand functions for standard consumer goods are typically not assumed to be linear, and that this is attributable largely to one well-known phenomenon observed in conventional markets: diminishing marginal utility. The fifth can of soup I purchase *might* bring me about as much "satisfaction" as the first, but it is not likely that we would anticipate this of the fifteenth. This is because, economists tell us, the marginal utility of commodities we consume tends to decrease with their consumption, such that the next unit will bring somewhat less satisfaction than did the last one. While there is an overall decrease in utility, however, the rate of decrease is not presumed to be constant; the difference in "satisfaction" between the first two units consumed should be smaller than the difference in the last two units consumed. Thus an additional unit of a good while one has little of it is likely to be more valuable to that person than an additional unit when one already has a lot of it. Why, the economist might well ask, can we not say this of environmental goods as well? Rendering one lake in my area inhabitable for fish is likely to be no small matter, and may thus be significantly more valuable to me (especially if I like to fish) than the money I would save by withholding my contribution to cleanup efforts (or the admission fee to a newly privatized lakefront). Doing this for even more lakes may have additional value to me, to the extent that I might be willing to pay to keep these inhabitable for fish, but it's conceivable that this value may not have the same magnitude as that of the initial lake. There is, after all, only so much enjoyment I can get from these lakes, and much of it is likely to be associated with the first. Hence the same diminishing marginal utility in operation when I consume cans of soup may be in operation when I "consume" environmental amenities such as an unpolluted lake. Were an economist to probe my environmental preferences with a CVM survey, I might accurately "price" the cleanup of several lakes only slightly higher than the cleanup of the first lake. Despite our potential concerns about "embedding," there would be nothing particularly anomalous in these WTP assessments.

There may be a snag in this reasoning, however. Diminishing marginal utility refers to—quite literally—*utility*. The satisfaction we derive from things we consume for *use*, the claim goes, tends to diminish with additional units of the good. CVM surveys, on the other hand, are intended to capture WTP assessments for non-use values as well. As we have seen in the case of environmental goods, this includes the *option* value to use a good at a later time, the *bequest* value in desiring that others may make use of the good, and, perhaps most important for environmental concerns, the *existence* value in an environmental good's not being used at all. While it seems reasonable that phenomena such as diminishing marginal utility observed in the consumption of standard commodities might be generalized to environmental goods which seem to involve "use"—be it for purposes of recreation or survival (such as fishing for enjoyment versus fishing for a living) or whatever reason—it might not be clear how we would attribute the embedding effect to diminishing marginal utility when *use* of the good in question is not at issue. One can "fill up" fairly quickly on wilderness camping no less than one can "fill up" on butter-pecan ice cream or lawn furniture; hence we might speculate that the WTP value of the fourth week of wilderness camping is somewhat less than the first for many individuals, and that a four-week camping trip may be valued at less than four times a one-week trip. Yet it's not obvious that we may attribute this same property to the assessed value of wilderness *simpliciter*—without the provisions for camping or other recreational use. Thus we might object that, superficially at least, those environmental amenities that we don't actually intend to "use" in any meaningful sense should not exhibit diminishing marginal utility, and hence should not exhibit the embedding effect when "priced" in CVM surveys; the fact that they do should thus rightly be a cause for concern for environmental economists.[2]

I don't think, however, that the CVM practitioner would be willing to surrender the case just yet. Despite the worries described above, the case might be made that some diminution in the value environmental goods—even those valued on the basis of their existence alone—might be anticipated as more and more of the good is provided. Indeed, it might be argued that unless one is willing to believe that individuals would provide sincere WTP assessments for a virtually infinite allocation of wilderness, clean air, and the like, we would expect to see some degree of what appears as "embedding." If I really only desire to see half of the wilderness area in a particular region preserved, and I'm willing to pay $x for this amenity for the sake of wilderness preservation alone, one would be hard pressed to get me to say that I would be willing to pay $2x to see *all* of the wilderness area preserved (unless I was confused about the question or I tended to up my bid to provide a more "socially desirable" response), and one would have little reason to anticipate a linear extension of my WTP assessments. You might be able to persuade me that I'm willing to pay a little bit more than $x to preserve 65 percent of the wilderness area in the region, and perhaps even a bit more than that to preserve 75 percent. It's likely, however, that if such WTP increments existed at all, they would get increasingly smaller—even if not in the smooth function of diminishing marginal utility. It's also likely that there is a point at which I would pay no more for any marginal increase in the amount of wilderness area preserved. Regardless of the differences we might observe between environmental goods (described as public goods or not) and typical commodities, we might reasonably anticipate that the amounts individuals desire of the former would be no less finite than the amounts they desire of the latter, even if it takes a seemingly bizarre survey question to prompt those individuals to think about the environment in such terms. While they may talk glowingly of "unlimited wilderness" and "clean air at any cost" in daily discourse, if we believe that CVM surveys are tapping anything reliable and useful in the way individuals value the environment, we should probably believe that one who cooperates enough to offer a sincere WTP assessment must have an imagined "ceiling" on the allocation of the environmental good in question—one above which she would not be willing to pay for any marginal increase. And if this were true, it would seem to be no less so for non-use values as it is for use values. While WTP amounts in such cases need not approximate the diminishing marginal utility curves of twentieth-century economics, the fact that there are limits to what one would be willing to pay for certain amounts of environmental goods suggests that "embedding" should not be all that surprising.[3]

I've been enumerating here a number of reasons why we might acknowledge the embedding effect apparent in CVM survey results and say "So what? Why should we believe that this estranges CVM from orthodox economic theory or that it should strike one as entirely counterintuitive?" There are, however, other grounds on which we might infer that an observed embedding effect in CVM surveys would cast doubt on the validity and reliability of WTP assessments as measures of the economic value an individual places on the environment. In an article that has come to invite a substantial amount of attention from both proponents and critics of CVM studies, Daniel Kahneman and Jack Knetsch (1992a) argue that the embedding effect is attributable to the fact that CVM surveys are not assessing the economic value of an environmental good but rather the moral satisfaction which contributions to it bring. It is not, they suggest, the subjective economic value of the good that one affirms with a stated WTP but rather the "symbolic" value: I'm not just "purchasing" the survival of a particular species or a river, I'm attesting to the environmental preservation of species and rivers in general. CVM thus measures not "the utility derived from increasing the total supply of the good" but the "utility gained in the act of giving" (Kahneman and Knetsch, 1992a:64). Given this, they argue, the observed embedding effect makes sense: if individual respondents see themselves as purchasing moral satisfaction in offering a WTP bid, varying levels of the good in question should not make all that much difference in their WTP. Since it is the good's symbolic value as an object of moral concern that one is assessing, providing more or less of it should be largely irrelevant in what one is willing to pay to see it provided. Kahneman and Knetsch's own empirical studies appear to support their arguments. Using a CVM instrument that assessed both perceived "satisfaction" (on an ordinal point scale) in contributing to more and less inclusive forms of a dozen public goods (about half of which were environment-related) as well as WTP assessments for each, the researchers found a high rank-order correlation between the satisfaction reports and median WTP, as well as strong evidence of embedding (bids for less inclusive goods were not substantially different from those for more inclusive versions of the good). Kahneman and Knetsch conclude from this that it is moral satisfaction that exhibits embedding, not the perceived "economic value" of the good: "if the inclusiveness of the cause does little to enhance moral satisfaction, increasing the inclusiveness should have little effect on WTP" (Kahneman and Knetsch, 1992a:67). It is the perceived moral satisfaction for contributing that, the claim goes, has primacy here. Because *it* exhibits no measurable increase for more inclusive versions of a good, neither does the WTP assessment for that good.

Are the conclusions about "moral satisfaction" which Kahneman and Knetsch draw reasonable? While a number of methodological concerns might arise over the ways in which they collected and assessed their data,[4] at least one potential problem suggests that their hypotheses were ill-formed from the start. If WTP assessments are in fact an indicator of "charity"—that is, if they are measuring the "warm glow" obtained in contributing to what are perceived as "good causes"—we would expect to see similarities not only between more and less inclusive forms of a particular good but also between *different types of goods*. Since what we are "purchasing" is a flat rate of moral satisfaction presumably not tied to discrete quantities of the good itself, we might anticipate no substantial differences between WTP for protecting the peregrine falcon and reducing damage from acid rain in Muskoka, Ontario (or, more inclusively, protecting all endangered birds and reducing the effects of acid rain in Ontario at large—all of which items in Kahneman and Knetsch's original CVM survey); it is a psychological need being fulfilled by "giving" to theses "causes" that is assessed, not the utility in having more falcons in the world or fewer sulfur deposits in local vegetation. Yet we *do* see differences between the WTP assessments for these goods, in both their less and more inclusive forms, and these differences do not seem to be cause for concern for Kahneman and Knetsch.[5] Should these differences not suggest, however, that it remains something in the perceived value of the good itself (economic or otherwise) to the individual that is being assessed, and not a flat-rate "moral satisfaction"?

Surely Kahneman, Knetch, and anyone else convinced that WTP measures for environmental amenities are mapping moral sentiments rather than strictly economic valuation would protest here. "Of course we would expect to see differences in WTP *across* different goods in spite if the absence of any substantial difference between WTP *within* more and less inclusive forms of the good. It's perfectly conceivable that I might consider saving species of birds more important morally than reducing the damage from acid rain regardless of the size of the area benefitted; just as people have different economic preferences and priorities, they may have different moral preferences and priorities. We might thus anticipate that if my 'moral satisfaction' in contributing to bird preservation is greater than that for contributing to acid rain clean-up, we would expect my willingness to pay for these efforts to be similarly greater—this in spite of the fact that I'd be willing to pay no less for preserving a single species of bird or ridding a very local area of the effects of acid rain." Yet the question remains: If we can make such metric extrapolations in our moral satisfaction and WTP *across* various "causes," why is it we don't—according to Kahneman and Knetsch's inferences from their own data—make them *within* various levels of the goods provided by such "causes"? Why is it that the faculties which allow us to make categorical distinctions between different types of environmental goods and the "price" we are willing to pay for them break down when we are asked to assess varying amounts of those goods? If we are astute enough to understand our preference for apples to oranges—and our willingness to pay more for apples—as a matter of the moral satisfaction that promoting apples provides, why are we not astute enough to prefer a bigger apple to a smaller one? Just what is being "explained" here?

Rather than accounting for the appearance of embedding in WTP by reducing it to what they take to be a more fundamental "moral satisfaction" that comes with contributing to environmental goods, as they purport to do, Kahneman and Knetsch have merely prompted the further question of why we have the appearance of embedding in moral satisfaction itself. The existence of constant "moral satisfaction" levels within more- and less-inclusive levels of an environmental good doesn't so much explain embedding in WTP bids for those goods as it leads us to wonder why there exists a constant level of "moral satisfaction" within levels of the good in the first place and in spite of the fact that people apparently and not surprisingly can derive more moral satisfaction from contributing to some goods over others. If we thought it odd that someone might pay no more to preserve all species of birds than she would pay to preserve peregrine falcons, why would we think it any less odd that she experiences as much "moral satisfaction" in the latter as in the former? The fact that she might derive greater "moral satisfaction" from reducing acid rain damage in Muskoka, Ontario tells us that she is able to draw moral distinctions and prioritize those causes she deems more important (and adjust her WTP accordingly), yet we are apparently to believe that her sensibilities fall apart when it comes to her assessment of different amounts of the same good—both "moral" and economic.

Why am I concerned here amid a philosophical examination of the moral duties of environmentalism with what no doubt sounds like methodological ephemera in a particular economic survey instrument? Let's review what we've seen in this brief chapter—why, despite seemingly plausible explanations for it, "embedding" in CVM surveys seems to matter—and consider where this is leading us in Part II. We observe a correlation between measures of WTP and "moral satisfaction" for a number of public goods and we are drawn to attribute the more fundamental, if not causal role to the latter. What we are really addressing here concerns not the reliability of CVM surveys but rather their validity: it's not one's economic preferences for which WTP serves as an indicator, but rather the gratification one has in fulfilling what one believes to be a moral obligation. WTP may be a measurable economic phenomenon, but it is merely epiphenomenal, this concern suggests, to the moral satisfaction that is occurring in bidding on a "good." It is precisely because "moral satisfaction" exhibits some degree of embedding while it correlates with WTP that Kahneman and Knetsch feel justified in attributing to it its more fundamental role; if, "moral satisfaction" tended to vary within the goods' levels of inclusiveness while WTP remained the same, we presumably would not be able to draw the conclusions that they do.

Even if the authors' case were a strong one, and it were indeed true that the reported WTP for environmental goods is in fact one's willingness to pay to achieve "moral satisfaction" in one's treatment of the environment, this by itself wouldn't necessarily undermine the CVM. This is precisely the problem. To the extent that CVM can accommodate the claim that what is being measured is in fact moral approval of seeing an environmental good procured, preserved, etc., this suggests that our moral sensibilities have a shadow "price" no less than our consumer preferences. Kahneman and Knetsch have merely drawn attention to a more fundamental set of normative questions concerning the relation, if any, of our moral commitments to our consumer preferences, whether the former can be tapped as if they were the same in nature as the latter, and what this all means for our attempts to determine our priorities in enacting environmental policy. We turn to these concerns in the remaining chapters.

NOTES

1. Kahneman and Knetsch (1992b) describe these as forms of *perfect* embedding, to distinguish them from what they call *regular* embedding, wherein WTP assessments for a good vary depending on whether it is offered in isolation or as a bundle of other goods. By this account, *perfect* and *regular* embedding appear to be the same phenomenon detected via different means: with the former, we vary the scope or quantity of the good and note that WTP remains roughly constant; with the latter, the scope or quantity of the good remains constant, but we observe the same WTP for both the good in isolation and as part of a package of goods, and infer that the WTP for the same good is lower when it is included in the package. For example, one's WTP of $100 to save a single herd of African elephants would exhibit *perfect* embedding, by Kahneman and Knetsch's account, if the same individual's WTP to save the entire species was also $100. If, moreover, her WTP to save a herd of African elephants *and* to restore orangutan populations in Borneo was also $100, she would be exhibiting *regular* embedding. Clearly the same phenomenon is at work here; either the environmental good in question appears to lose value between the situation where it is taken in isolation and the situation in which it is considered among other things (or more of the same things), or the marginal value of the additional goods is inexplicably small.

2. To extend the argument: even some cases where the environmental good may reasonably be said to be "used" might strain the plausibility of attributing embedding

to diminishing marginal utility. Consider, for example, the case of clean air. Do we really want to say that an individual who would be willing to pay the same amount for a 15 percent reduction in particle air pollution as she would for a 5 percent reduction is experiencing the diminishing marginal utility of breathing?

3. Some defenders of CVM—Harrison (1992:249), for example—suggest a different phenomenon consistent with orthodox value theory accounts for the embedding effect: substitutability. The demand for two goods that are perfect substitutes, according to conventional economic understanding, is inversely proportional; the more we have of one, the less we want of the other. Thus if environmental good Y is a perfect substitute for environmental good X, we might reasonably expect that the WTP for environmental "package" Z—which merely "bundles" X and Y—would be no more than the WTP for X or Y alone. There is nothing here, Harrison and other advocates of CVM claim, that is inconsistent with what we already understand about economic behavior with regard to goods in markets. As plausible as this sounds, however, we should keep in mind that substitutability can only account for the "embedding effect" in those cases where the more inclusive good in question combines different, discrete environmental goods. In situations where the embedding appears to be categorical, temporal, or of a linear extension of the less inclusive good, its applicability seems strained. It would be difficult to imagine a situation where we might interpret "all wolves minus timber wolves" as a perfect substitute for timber wolves, or 99 of them as a perfect substitute for one, or an additional five years to save them as a perfect substitute for the first five years. Kahneman and Knetsch (1992), moreover, in the article to which Harrison is responding, had already anticipated the possible interpretation of embedding as a manifestation of substitutability. They quite agreeably point out that while substitutability may account for the appearance of an "embedding effect" with the *use* values attached to environmental (or any public) goods, "the uniqueness of the valued good is the essence of existence value" (Kahneman and Knetsch, 1992:59). Hence to suggest that existence values can be perfect substitutes for each other can only make sense if we abandon our understanding of what existence values are. This may or may not be an attractive option, but it would need further justification.

4. The two most immediate responses to Kahneman and Knetsch's article—V. K. Smith (1992; see also the authors' rejoinder, 1992b) and Harrison (1992)—both focus more on the methodological shortcoming of the authors' survey and thus on the question of whether they actually detected true embedding and not their conclusions concerning the moral content of WTP value assessments. Harrison, however, quite rightly points out that Kahneman and Knetsch's belief that they were measuring "moral satisfaction" is optimistic at best, given that the term "moral" is never invoked in the "satisfaction" question that they employ. While Harrison concludes from this that much of their discussion on the "purchase of moral satisfaction" can be dismissed as nonsense, I give the authors the benefit of the doubt and assume, for the sake of argument, that they have good reason to believe respondents interpreted the instructions to ". . . Indicate the degree of satisfaction you would receive from contributing to each cause on a scale 1 to 10 . . ." (1992a:66) as particularly *moral* satisfaction.

5. Interestingly, we don't see—impressionistically at least—a substantial difference in the mean "moral satisfaction" in contributing to the two goods mentioned

(7.18 versus 6.46 for the less inclusive versions of the good, and 7.25 versus 6.98 for the more inclusive versions; Kahneman and Knetsch, 1992a:65). The authors do not test statistically the mean differences on the 10-point moral satisfaction scale between any of the goods included in the survey, though we might infer from the reportedly high rank-order correlations between satisfaction and WTP that there are differences among values of the former.

Chapter 10

The Rationality of Environmental "Charity"

If it appeared that people who donate money, time, or other resources to environmental causes behaved "rationally" and "economically" in doing so—say, by calculating their charitable actions for maximum impact rather than to merely feel good about themselves—would this count as evidence that we do in fact "price" the natural environment and the "goods" it provides like we do any other item allocated on a market? Or would it merely suggest that our understanding of moral duties is no less reasonable than our economic preferences? How we might answer these questions determines which direction we might go with the perceived "contamination" of WTP measures of environmental value by our moral sensibilities.

There are at least two ways to interpret the claim, raised in the last chapter, that to put an economic value on the natural environment (or on some specific aspect of it) is to permit one's moral values to influence somewhat one's economic preferences. One might suggest that this distorts our stated economic preferences, or one might suggest the seeming confluence of the two is actually illusory, and that are moral values and economic preference in fact do not "mix." The first is a concern of the free-market economist bent on locating the true "value" of the natural environment in the eyes of its "consumers"; the latter interpretation might be held by environmentalists who, understandably, would adhere to the objection that we are asking the wrong question when we ask someone to put a dollar amount on the "value" of the environment to her. Given their importance to our pursuits here, let's state these two positions a little more fully and precisely:

I. One might hold that the determination of the "value" of the natural world should indeed be based upon the subjective value that people place on it (even that it can be economically "priced") but that one's stated, monetized preference for environmental amenities is often not a reflection of what one desires but rather what one feels one *ought* to pay to preserve species, cleanup the effects of pollution, etc. This renders, one might further believe, one's WTP assessments less valid and thus less useful in determining the aggregate "value" of the environment. Note that the claim here is not merely that this "warm glow" effect systematically biases WTP assessments for environmental goods upwards, thus inflating the perceived value of the environment beyond what people would actually be prepared to pay.[1] The claim here is a stronger one: in so far as *any* moral motivations influence WTP, the resulting values cannot be considered commensurable with orthodox economic interpretations of "value." CVM thus involves the inherent *mismeasurement* of what individuals might actually be willing to pay for environmental goods; moral values can only confound economic preferences.

II. One might reject, on the other hand, as do Sagoff (1988a,b) and a number of other environmental theorists, the notion that the natural environment can be "priced" by subjective assessments of "economic value" at all. The quality of the environment is, this claim suggests, rather than the object of economic preferences, a concern of one's normative values—values which cannot be directly compared with the preferences typically exhibited by consumer behavior in conventional market situations. It is no doubt conceivable, this claim implies, that economic preferences can be traded against each other, and even that individuals must occasionally weigh their value commitments against each other and "trade" certain "amounts" of one for another as if they behaved like economic preferences (as we often do in moral dilemmas). CVM surveys, however, as with any attempt to put a price on the natural world, require that individuals consider their values as if they just were the *same thing* as preferences and reduce both to the common measure of money. Thus where Kahneman, Knetsch, and other economists dismiss CVM as prone to mismeasurement to the extent that it cannot insulate one's WTP assessments from one's moral inclinations, Sagoff and other environmentalists decry it as prone to missing the normative content in one's valuation of the environment entirely, and thus bound to commit, by design, a *category mistake*.

In this chapter and in the next, I will argue that neither of these constitute the primary normative concerns we ought to have about CVM or any attempt to base the value of the natural environment in social policy on the exclusively subjective value of environmental goods. The next chapter addresses (II) above—the question of whether soliciting one's "willingness to pay" for environmental goods commits a category mistake; in this chapter, we address (I).

Kahneman and Knetsch, as well as other critics of CVM (e.g., Diamond and Hausman, 1993:27–29; Milgrom, 1993), clearly hold position (I) above—one which effectively asserts that CVM is misguided insofar as what it solicits is uncharacteristic of what conventional economic value theory would say about our value assessments of consumer goods in real markets (as opposed to the hypothetical markets CVM is intended to tap). We pay precisely what we are willing for cans of soup (provided some producer is willing to make them available at that price), not what we feel we "ought" to pay, and if someone were to ask us what we might pay for that can in a hypothetical market, we are not typically inclined to reflect upon any overriding moral obligations concerning the provision of soup. True markets would be disrupted if individuals priced things out of a sense of "duty"; for public goods such as environmental goods to be "priced" via the aggregation of WTP measures, Milgrom affirms, "It would be necessary for people's individual existence values to reflect only their own personal economic motives and not altruistic motives or sense of duty, or moral obligation" (Milgrom, 1993:431).[2]

But is this necessarily the way we understand pricing in consumer markets, as if void of any motivations besides self-regarding "preferences"? Arguably not. What these critics of CVM are implying is the claim that economic preferences are necessarily governed by egoistic motivations, while our moral values are uniformly altruistic. These are, we are presumably to infer, separate domains which should never overlap. Whether such advice is wise in itself, it certainly doesn't seem to be a guiding principle in conventional welfare-economic accounts of "value." Such accounts do not require that *psychological egoism* be true in order for them to be coherent. It is not necessary, in other words, to believe that individuals act primarily out of self-regard to believe that they can make reasonable assessments of what is important to them and that social policy should somehow be attached to those assessments. *Egoism* refers to a particular kind of *justification* for doing certain actions or holding certain values: such actions or values are considered *justifiable* insofar as they further the interest of the self. That Hans is an agent doing action A or holding value V is not sufficient to claim that A and V are egoistically motivated; neither would be the fact that A or V may further the interests of Hans. What matters in telling us whether A or V are *egoistic* is not the identity of the agent or the beneficiaries of his actions or values. What matters is whether Hans does A or holds V *because* it would further his interests. Not all of Hans' preferences will meet this condition. He may prefer to do some things that will benefit his family or friends, perhaps at his own expense, or that will further the interests of society at large, or even the natural world. He will even likely do things and hold values that further his own interests but for which his justification refers sincerely to something or someone else. The objection typically offered at this point—that even when

Hans is acting on behalf of others, he is gratifying one's own desire to act altruistically—merely reaffirms the true-but-trivial fact that one's desires are one's own, whatever the object of those desires may be, and does nothing toward distinguishing the types of motivations we might have (since all of our decisions on the actions we take and the values we hold are ultimately our own). Egoistic preferences and desires are aimed at the self and are justified to the extent that they benefit the self; they are a particular kind of preference, and no reasonable interpretation of what economists have to say about "value" would require that we consider only this narrow band of desires in an economic accounting of what individuals want.[3]

By the same token, while conventional economic accounts of value do not require that individuals be egoistically motivated, it's also true that they do not preclude the possibility that individuals are motivated by purely altruistic or other moral sentiments. We don't, as W. Michael Hanemann (1995:105) aptly points out, ignore Catholics when estimating the demand for fish, or make adjustments when childless couples vote for school bond issues. For the same reasons this strikes us as appropriate, economists can claim, we do not need to discount WTP assessments for environmental goods that are based on apparently selfless inclinations. Preferences, as welfare economic and social choice theories understand them, are simply preferences; it is not necessary that we discount them when they happen to be based on the well-being of others.[4] Occasionally this point gets clouded when we claim to be acting on consumer "principles," rather than "cold, economic calculation"—as when one purchases detergents or hygiene products at a premium for their avoidance of animal testing, or when one pays a higher price for organic produce than conventional produce purely in an attempt to support a fledgling but morally commendable industry. Yet these are consumer preferences upon which we act no less than the desire to pay the lower price for rabbit-tested shampoo and agribusiness carrots. The fact that we make such decisions on the basis of moral or political principles need not render such decisions "irrational" or outside the bounds of what we understand as consumer demand. In fact, what renders such purchases at all "rational" in an economic sense is the fact that we are assessing something more than the good itself when we make such decisions—something perhaps less tangible than shampoo or carrots, but something which accords with particular desires we hold and can apparently be "priced" no less than other goods. Kahneman and Knetsch tap this in hypothetical environmental "purchases" and conclude that it undermines one's attempts to assess the subjective value of environmental goods to the extent that purchasing "moral satisfaction" is not something rational economic agents do. Such a conclusion seems arbitrary, and no more plausible than the conclusion that it is precisely *because* we can weigh simultaneously both monetary costs and moral principles in our purchasing decisions that we are "rational" consumers, and that it is one of the tasks of economics to quantify the demand for things given that people may prefer them for a variety of reasons. Whatever we may think of consumerism or the rationality of an economic system which promotes it, that moral inclinations can weigh in on our consumer decisions doesn't seem at all incommensurable with the way contemporary economists understand demand and their role in assessing it. Welfare economics and its counterparts may be accused of reducing the complexity of human motivation to what amounts to context-specific market exchanges, but such theories are not so naive as to believe that we are simply calculative automatons, eschewing principles, traditions, and the counsel of others for the "best price" in the pursuit of our desires.

There is room here, however, for a further argument on behalf of the concerns expressed in (I) above. Even if, one might assert, conventional accounts of value see no need to distill our economic motivations, pulling out the selfless, altruistic, and other moral inclinations and leaving only the egoistic ones, one might argue that overtly altruistic "purchases"—in their most common form of the charitable contribution—behave differently from conventional consumer purchases and should thus not be conflated with the latter. Such an argument would posit that even if conventional economic theory (as well as CVM) acknowledges no difference between egoistic and altruistic motivations in economic activity, we have empirical evidence which suggests that they *ought* to be treated differently. Diamond and Hausman (1993:29) suggest (quite incidentally), for example, that charitable contributions from individuals do not exhibit the characteristics we see in their "rational" consumer purchases. The *reconstructed argument*—let's call it that, for reference purposes—approximates a fairly straightforward *modus tollens* format:

1. If charitable contributions were of the same nature as economic "bids," we would expect people to contribute to charities as rational consumers of what the charities provide—that is, rather than giving small amounts to many related causes (where the impact of each contribution is likely to be negligible), they would consolidate their contributions for maximum impact in one, or, at most, a small handful of charitable causes.
2. People do not behave as rational consumers when contributing to charities (they do in fact tend to disperse their donations across several causes).

Therefore

3. Charitable contributions are of a different nature from consumer purchases

Consumer purchases, in short, adhere to at least some of the prescriptions we might have for rational consumer behavior, in terms of "getting the most for one's buck"; charitable contributions do not exhibit these characteristics (regardless of whether we might expect them to) and hence seem to define themselves as a different type of activity. The substantive implications of this argument, if valid, would be clear for CVM and any attempt to solicit from individuals a subjective value assessment of environmental goods: if charitable contributions exhibit different, non-rational characteristics from those we see in individuals purchasing goods on a market, it would be dangerous to risk describing as true WTP "bids" those CVM responses that are motivated by altruism, a sense of duty, or interpretable only as one's "willingness to contribute to an environmental, charitable cause." If the utility of charitable contributions is in the giving rather than in the "good" provided, CVM would indeed not be assessing potential demand for environmental goods in a hypothetical market but rather the inclination to endorse environmental "causes"—two very different things.

The argument has an undeniable ring of plausibility to it, due largely to the tendency for the second, empirical premise to ring true to our experiences. Charitable contributions—including those offered to environmental organizations—tend to be spread around by individuals in small parcels (relative to one's total budget for "charity"); we give a little to Greenpeace, a little to the World Wildlife Fund, a little to the efforts of a local recycling organization, etc. (That the premise elicits only our impressionistic assent is not a major problem. It remains an empirical hypothesis—one that we could test if we chose, and can be accepted without evidence at no loss to this discussion). Few of us are likely to have reasoned while writing out the checks or filling out the online form that, rather than giving $40 to each of five disparate environmental organizations, we would be wiser to give $200 to one of the organizations, to give our charitable inclinations more leverage in affecting environmental quality.[5] Yet determining which of those (or any) organizations should receive the consolidated $200 is precisely the problem here, and points to the weaknesses in the first premise in the reconstructed argument. I could, for example, donate the consolidated $200 to the charitable organization which strikes me as the most important, or is on some other grounds my "favorite." This would be a decision based entirely on my fancy, however, and not on the grounds of an economic "rationality." Moreover, it wouldn't necessarily be more economically effective than splitting my contributions among five charities. It could well be that my favorite charity is also the favorite of many other people, each of whom has "rationally" opted to donate their available resources to this single cause than to split their contributions across several causes. My consolidated donation would thus have less of an impact on this cause than the reconstructed argument would suggest; I've pulled resources away from causes less popular and which would thus get a bigger boost from my donation and diverted them toward a cause where their marginal impact will be relatively small.

Suppose, then, that we ignore our "favorite" and donate all the disposable funds we have designated for charitable distribution to the cause for which the contribution would have its greatest impact—the one that the reconstructed argument might describe as the economically more "rational" object of our charitable inclinations. This is likely to be the cause which has the fewest resources since, all other things being equal, our contributions will have a more substantial marginal impact with this cause than with one which already has a lot of money. There may be, however, good reasons why this charitable cause has fewer resources than its counterparts. Imagine, for example, that one is choosing between contributing to a cause which aims to restore ocean habitats, one which will reduce local sulfur dioxide emissions, and one which will "save the cockroaches." Common sense suggests the last one will be the least well-funded from among these, and hence would be the one for which one's contribution would have greatest impact—possibly doubling the foundation's assets with a single contribution of a few dollars. Do we want to describe as "rational consumer behavior" a contribution to the Save the Cockroach Foundation on these grounds?[6]

The problem in the reconstructed argument here seems to be in the conditional statement that comprises the first premise. The claim is essentially that if charitable contributions were of the same "rational" nature as typical consumer activity, we would see particular patterns of behavior in the way people donate resources to charitable causes. Since we don't see such patterns of behavior in those contributions, the authors conclude, charitable inclinations and consumer preferences must be two distinct entities. Yet if we take the patterns of behavior that Diamond and Hausman would expect to see to their logical ends, we don't envision activity that is likely to strike us as "rational." In the reconstructed argument's counterfactual world in which charitable contributions are orchestrated like ideal market transactions, we would expect "under-funded" charities—presumably "under-funded," in part, by virtue of their unpopularity—to be bolstered by "rational" benefactors until some charity equilibrium is reached and the marginal impact of one's contribution would be equal across causes. This doesn't seem reasonable, and would have us giving away our money to absurd and otherwise neglected causes on the basis of how well our contribution would help them achieve their particular goals rather than on the basis of how well their goals coincide with our desires about the world. While the relative impact of one's contribution may have *something* to do with how we allocate our funds to charitable causes in a "rational" way (we don't like to think we're "throwing our money away"), this doesn't seem to be the best argument available to demonstrate that charitable contributions tend toward irrational forms of allocation.

We might, however, be able to salvage Diamond and Hausman's intent in this reconstructed argument by constructing a different scenario—one where the altruist weighs not competing charities but rather the option of contributing to a pooled charitable cause with other individuals against using her resources to bring about directly an improvement in the welfare of others. It might be possible to argue not only that "rational" altruists, when faced with such allocative decisions, tend toward irrational decisions, but that such decisions are inherently irrational—perhaps even incapable of being made. Surely this would, if demonstrable, lend some credence to argument (I) sketched in the opening sections of this chapter, and validates the fear that altruistic or other moral inclinations regarding the natural environment distort WTP measures and the presumed economic valuation of environmental "goods."

Before we pursue this, let's clarify the language we are using. What Diamond and Hausman presuppose and what I've built into our "reconstructed argument" bent on evaluating their concerns is what moral philosophers generally refer to as a *principle of rational beneficence*: a rule which would instruct one to maximize the benefits one can bestow upon other people or states of affairs by maximizing the positive difference she herself makes.[7] Defining a coherent form of such a principle takes up much of Derek Parfit's concern in the early chapters of his monumental *Reasons and Persons* (1984), and, as Liam Murphy (1993) points out, amounts to simple consequentialism minus self-regard (though, like consequentialism, it ignores any deontological restrictions that might be at issue). Offering one's only warm coat to another person, for example, may be considered commendable no matter who the recipient is or the circumstances of the offer. It would require a further principle of the kind at issue here in order to assert that it is *better* in some meaningful normative sense to offer the coat to someone who currently has no protection against the elements than to someone who is bedecked in an expensive down-filled parka. Thus a principle of rational beneficence is intended to serve as a decision rule for the altruist concerned about the most efficient way to allocate the benefits she has to offer.[8] Diamond and Hausman, it appears, want to argue that individuals contributing to charities tend not to adhere to such a principle—that is, that other considerations undermine whatever intentions they may have to maximize the good that they can do—and that this is not characteristic of their consumer activity in market situations, where they do tend to maximize the benefits they themselves can reap. As we have seen, my "reconstructed argument" detailing their point fails irrespective of whether or not we test their empirical second premise about what individuals actually do; the consequent in the conditional statement which comprises the first premise projects a situation that can hardly be described as "rational," and would, if achieved, do nothing toward fulfilling any principle of rational beneficence.

Regardless of whether the reconstructed argument could be made to work, it might be more effective to argue *not* that individuals tend not to adhere to a principle of rational beneficence when they allocate funds to charitable causes but that they *cannot* do so, and that charitable contributions of the type described here are thus inherently non-rational (and should thus not be conflated with consumer preferences). Note that the principle says nothing about the actions of other rational altruists who might be similarly inclined and who might thus be similarly bound by the principle; as stated here, it concerns only the maximum good that one can accomplish oneself. Yet, as Diamond and Hausman imply, rationally allocating one's resources among charitable causes must have something to do with how others are making similar decisions. Could this fact make for a more compelling argument on behalf of the concerns expressed in (I), outlined at the beginning of this chapter?

Consider the following scenario:

> The Nature Institute (NI) is an environmental advocacy organization. If every adult in the United States contributed 25 dollars to the NI, the organization would be able to set in motion, via worldwide legislative lobbying, mediation, and the creation of various market incentives, a series of policies by which the most serious threats of climate change could be significantly reduced if not averted over the next 50 years. No single individual's contribution to the NI would make an appreciable difference in the total amount collected, or in the corresponding ability of the NI to bring about environmental policies. It would require, however, the contribution of nearly two hundred million individuals for its impact to be felt.
>
> Millhouse has 25 dollars to contribute toward such a cause. Millhouse is also aware of the NI and of its efforts—and of its claims as to what it could do provided a fairly high level of cooperation. He also, however, has made the acquaintance of a homeless man in his neighborhood, one who cheerfully greets him every day and sells him a newspaper once a week. Millhouse know that the man lives hand to mouth and rarely enjoys a good meal. The 25 dollars that Millhouse has earmarked for the NI, he reasons, could go a long way toward alleviating this homeless man's misery, buying him the most sumptuous meal he's had in years.[9]

The principle of rational beneficence would seem, in this scenario, to instruct Millhouse to withhold his contribution to the NI and instead use his 25 dollars to purchase a sumptuous meal for the homeless man. If Millhouse is to maximize the difference *he* can make—if he is to bring about the most good that *he* can affect—he would seem obliged to provide the great deal of joy that a lavish meal would bring to the homeless man, since this good would seem to outweigh whatever marginal difference Millhouse would make toward averting climate-change catastrophes by contributing to the NI.

If this were true for Millhouse, however, this would be true for everyone else. The principal of rational beneficence would seem to require each individual to forego her donation to the NI and instead direct her charitable contribution toward a non-cooperative venture such as purchasing a meal for a homeless person.[10] But in doing so, charitable organizations such as the NI would never thrive, and issues like climate change (or, for that matter, hunger less proximate to those with the resources to buy someone a lavish meal, as in many parts of South Asia, Latin America, and Africa) might never get addressed.

Does this suggest the disturbing conclusion that the principal of rational beneficence cannot coherently guide altruists in the allocation of their good will? If true, it would imply that charitable contributions—regardless of their size (for all but the immensely rich such as Microsoft founder Bill Gates) or of the cause to which they are applied—are misplaced attempts to benefit other persons (or things, as in the environment), and that voluntary contributions to cooperative schemes are inherently irrational. And if *this* were true, it might further imply, as Diamond and Hausman had attempted to argue, that there are fundamental differences in the nature of charitable contributions (which cannot be rational) and true market transactions (which, for a variety of reasons, can). The revised "reconstructed argument" would thus look like this:

1. Individual contributions to charitable causes always could have been allocated to non-cooperative ventures where their impact would have been greater

   Therefore

2. Charitable contributions could never adhere to the principle of rational beneficence
3. Private consumer purchases are rational insofar as one can maximize one's interests in a market

   Therefore

4. Charitable contributions are of a different nature than consumer purchases

Plausible as it might sound when laid out systematically like this, such an argument may be too hasty in pronouncing charitable contributions "irrational" simply because they do not conform to the principle of rational beneficence. One might concede the soundness of this version of the argument but object that there may be other reasons for contributing to charities that subscribe to coherent, generalizable moral principles. This in itself is not an indictment of the principle of rational beneficence, but rather a claim that it sometimes has us do things in conflict with commonsense morality or other moral principles and thus must be occasionally overridden.

Derek Parfit has himself argued, for example, that there are situations in which a rational altruist ought not to adhere to a principle of rational beneficence—specifically, when he could be a member of a group which brings about a greater amount of good than he himself can affect. This claim is captured in Parfit's "proposition C10"—what we will call here, taking a cue from Otsuka (1991), the *principal of group beneficence*:

> When (1) the best outcome would be the one in which people are benefitted most, and (2) each of the members of some group could act in a certain way, and (3) they would benefit these other people if *enough* of them acted in this way, and (4) they would benefit these people *most* if they *all* acted in this way, and (5) each of them both knows these facts and believes that enough of them will act in this way, then (6) each of them ought to act in this way. (Parfit, 1984:77)

If a principle such as that of *group beneficence* could be said to be more appropriate in guiding the altruist in questions of when and how to contribute his resources toward cooperative ventures such as charities, this might undermine the quality of premise (2) in our revised reconstructed argument, since the principle of rational beneficence would be irrelevant in such a case. Yet Millhouse might rightly wonder how this principle is to help him decide, since it still seems that, yes, while averting a global climate catastrophe may agreeably be considered a greater good than a single sumptuous meal for one homeless man, he himself can do more good by providing the latter rather than using the funds for his relatively meager contribution to the former. He seems to be presented with two alternatives, each of which would be justifiable according to different and, in this case, mutually incompatible principles. How is Millhouse, or anyone, when faced with the option to act altruistically via one's own endeavors or via a cooperative endeavor, to decide whether it is the principle of rational beneficence or the principle of group beneficence which is to justify one's actions?

It might seem on the surface that this is simply a case of competing and potentially conflicting moral principles and that there may not be reasonable grounds for deciding between them. The problem may be more profound, however. Since both principles appear to represent standard consequentialist inclinations, we might think that the one which should take precedence in a given situation is the one which is most beneficial. But in what Otsuka (1991) has aptly described as the "paradox of group beneficence," we may in fact not be able to pin down here an outcome which is "most beneficial."[11] For any outcome we might select, in other words, there would exist another outcome which is more beneficial. If true, this would surely reinforce our revised reconstructed argument and suggest that charitable contributions indeed cannot be "rational" in the same way market transactions can.

Consider the alternative outcomes in the situation of Millhouse and the other altruists with whom he might cooperate:

a. Everyone uses his or her $25 dollars to buy a sumptuous dinner for a homeless man
b. Some individuals buy a sumptuous dinner for a homeless man, others contribute to NI's efforts to avert global warming
c. Everyone contributes to NI's efforts to avert climate change

Outcome (a) is, it seems clear, less beneficial than outcome (c), since climate change isn't averted and everyone suffers the ill-effects (including all homeless people). Yet it also seems true, the paradox of beneficence suggests, that for every outcome other than (a), it is less beneficial than an outcome in which one fewer altruist contributed to NI and instead purchased a sumptuous meal for a homeless person. This would certainly be true of (c), since if *everyone* (N) were to contribute to NI, it would be better if N-1 had so that climate change could be averted and one homeless man could enjoy a meal. Yet if this were true for (c), it would seem to be true for any outcome captured by (b) as well—we could always have improved it by having one less NI contributor and thus one more meal purchaser, since it would seem that the particular agent in question is not going to make the difference in NI's efforts. But if this were true of everyone, all should purchase a meal for the homeless man, and we'd be back to outcome (a), with robust homeless people but with a global ecological crisis. Hence, we might be tempted to conclude, there is *no outcome* in this situation which would benefit most people, and neither the principle of rational beneficence *nor* the principle of group beneficence could be adhered to. Given that, at any particular time, the rational altruist is confronted both with options for contributing to group charitable efforts and to efforts in which she alone could benefit others, the paradox has fairly disturbing implications.

Is the paradox a genuine one, however? While it is superficially interesting, it should be clear—given the arguments in Chapters 1 and 2—that there is no "paradox" here, at least not in the sense that there is no outcome which would be most beneficial. For many of the ratios of meal-purchasers to NI-contributors that would constitute the array of possible (b) outcomes, it will indeed be the case that it would have been more beneficial had one less person contributed to NI and one more person purchased a meal for a homeless person.

This cannot, however, be true of *every* ratio, since there will be many of them in which the number of NI-contributors is greater than zero but less than the number needed to fund the organization's efforts in averting the effects of climate change. It would, as we can surmise from previous arguments, be difficult for any single contributor to know whether or not her contribution to NI would fall into this group (k, to revive the standard notation)—let alone whether hers would be the decisive contribution (i_k), but this fact alone is not sufficient to claim no such person i_k exists, or that one's membership among i_1 to i_k is absolutely indeterminate. An acorn is not an oak, a stone is not a heap of stones, and a handful of well-intentioned environmentalists is not an effective vehicle for staving off the effects of global warming; the difficulty in isolating a precise point in which one turns into the other is not alone adequate grounds for arguing that they are, in fact, indistinguishable. Given this, there indeed seems to be a range of outcomes other than (a) for which we could say it would *not* be less beneficial an outcome than one in which one fewer altruist contributed to NI (and instead purchased a sumptuous meal for a homeless man), since, if only k individuals were to contribute to the NI, one fewer among them would mean climate change is not averted, and this is worse for everyone.[12] Hence there is no paradox here, and while it may not be self-evident as to which of the two principles of beneficence at hand would benefit most people in such situations, we don't have a compelling justification for the claim that neither principle could.

What this discussion seems to be pointing to is that it is not the potential for confusing one's WTP bid with a charitable contribution that ought to be of concern to critics of CVM, but rather that such bids cannot be made in the economic vacuum that CVM presupposes. Prompting as to whether one's assessment will be treated as a market bid or a charitable contribution will cue one as to how payment would be conferred, but this doesn't do much to tell us who else will be paying and how much. The same problems which plague the rational altruist in allocating beneficence would plague the market participant in making WTP bids: contributing $x toward an environmental good, where it is only one contribution among many that would be necessary, is inevitably going to have less "value" than purchasing something which doesn't require the cooperation of others. This is little more than a re-statement of the problem of counterfinality we saw in Part I. When contributions are voluntary, it is difficult to meet condition 5) of the principle of group beneficence as we saw laid out by Parfit above. Our decision as to whether we could produce a greater benefit by contributing to a group effort or a non-group effort hinges on our perceptions about the rate of compliance by others in the potential group effort. But the payment is no less voluntary, and the decision no less difficult, when it is a market "bid" on a public good rather than a contribution to a charitable cause. Without the knowledge of how many others are likely to contribute toward the purchase of the good (and with the knowledge, given opportunities for "free riding," that they will each have an incentive not to), we are inclined to use our resources for non-cooperative consumer purchases and are thus in no better position than the altruist in securing the good. Hence it is not a problem of the "irrationality" of moral satisfaction or "warm glows" infiltrating otherwise rational market bids; the same problems of private and collective action affect market allocations of public goods no less than they do altruistic allocations to group charitable causes aimed at securing those goods.

One might suggest that the situation changes somewhat when a WTP bid is elicited as the amount one is willing to pay in increased taxes for environmental amenities; we can then assume that all (law-abiding) citizens will be contributing, and there is thus considerably less ambiguity as to what the rate of compliance is likely to be if one's bid were to actually affect tax policy and the fate of environmental goods. We will see later in Part II that this would do little to clarify one's WTP bid, since, while we would know everyone is required to pay *something* in tax schemes, we don't know what each would be *willing* to pay, and what *we* would be willing to pay would presumably need to reflect such assessments. More importantly, as I will also argue, a coherent assessment of environmental value—if we demand that it be expressed as "willingness to pay"—must presuppose a normative belief about what others *ought* to pay—a belief that is arguably not required for us to make purely charitable contributions.

"Charity," in short, in the sense held by advocates of position (I) outlined earlier in this chapter, does not seem to undermine CVM and the more general pursuit of subjective assessments of environmental value. People undoubtedly act "charitably" without regard to the potential effect or the contributions of others and thus in seemingly irrational ways—we give a dollar to the panhandler knowing full well that a dollar will go virtually nowhere toward reducing her misery and regardless of whether we believe others will be giving her a dollar as well. Such actions may indeed purchase little more than moral satisfaction and a "warm glow." That these actions are recognizable and perhaps common, however, does not establish that they constitute all of our acts of beneficence. It indeed seems possible that one can perform "charitable" acts on behalf of the environment under the presumed obligation of what one (but not necessarily others) ought to do and strictly for the "warm glow" that one might receive and yet still perform them more or less "rationally" with their maximal beneficial impact in mind. Hence to dismiss CVM and related methods of assessing one's WTP on the grounds that they elicit irrational, non-economic sentiments appears to be unfounded.

NOTES

1. Though some might suggest this as well; see, for example, Diamond and Hausman (1993:27–29).

2. One could probably argue along these same lines that aesthetic judgments should be similarly partitioned from assessments of economic value. Much in the same way that one's willingness to pay to prevent Van Gogh's *Starry Night* from being shredded and auctioned off as "souvenirs" would not reflect the "value" of the painting's beauty, the argument suggests, what we might be willing to pay to prevent a forest from being clear-cut and to preserve its beauty would not reflect the economic value of the forest (I thank Sandra Bartky for this observation and this telling analogy).

3. Recall that we touched on the heterogeneity of contemporary economic theory and psychological egoism briefly in the discussion of Adam Smith's "invisible hand" in the Introduction. Alan Holland—notably, a philosopher rather than an economist—presents this point nicely in regard to CVM, noting that the very "development of the concept of 'existence value' would seem to confirm" that all personal economic valuation is somehow egoistic (Holland, 1995:30).

4. The general point of interest here—that the ends of our actions, be they in maximizing our own well-being or that of others, are independent of whether or why we should choose the best means toward those ends—is as applicable to rational choice theory as it is to the particulars of economics. Choosing to effectively bring about our own ends is one thing; whether those ends substantively or consciously serve our own well-being (egoism), the well-being of others (altruism), or the well-being of all (utilitarianism) is another. David Schmidtz summarizes this point in a characteristically illuminating passage worth reproducing here:

> Some people see rational choice as essentially tied not only to a formal notion of means-end efficacy but also to a substantive notion of self-regard. However, I see rational choice's tie to self-regard as merely contingent. It so happens that people tend to have predominantly self-regarding ends. Since our ends tend to be substantially self-regarding, rational choice tends to be aimed at self-regarding ends, but that is not the same as thinking those or any other substantive ends are an essential part of the concept of rational choice. There might be ways in which morality is coextensive with rationality without being coextensive in the same way with self-regard. Nor is morality exclusively tied to other-regard. Just as there are rational reasons for other-regard, so, too, are there moral reasons for self-regard. (Schmidtz, 1995:23–24)

5. At the same time, however, few of us seem to contribute to charitable organizations willy-nilly and without regard for how much we are spending or the ends toward which the contributions are being spent. This in itself seems to undercut Hanemann's (1992) objection to Diamond and Hausman's more general case against CVM. Hanemann suggests that CVM respondents are not likely to confuse their bids with charity since "most appeals for charity are not designed to elicit one's maximum WTP—they just ask for some donation. They usually focus on the symbolic aspects of the donation, without identifying a specific consequence or an outcome that would not occur without it" (1992:90). While Hanemann's portrayal of charitable contributions may

have been plausible at some time, neither point rings true of contemporary charitable campaigns—especially those of environmental organizations. Solicitations are now often goal-oriented calls to action: particular legislative acts need to be defeated or passed, particular species need to be pulled from the brink of extinction, etc. It's hard to imagine, moreover, a mail solicitation from an environmental organization asking one to calculate the maximum one is inclined and able to donate and then contribute a lesser amount; they seem to be asking for the most one is willing to contribute. While the trappings of direct mail strategies may suggest otherwise (e.g., the array of $20, $50, $100, and $250 check boxes with a faux felt-tipped circle around $50 and a "personal" note from the executive director imploring one to "consider giving *at least* this amount), the goal of charitable solicitations—environmental or otherwise—still seems to be to get the most from each individual as possible. Whether or not this trend is actually dominant is, of course, yet another empirical question a bit beyond our scope here. Short of a representative sample of current charitable solicitations, however, the impressionistic evidence seems to suggest that they are not all that different from WTP solicitations in the ways Hanemann describes.

6. Diamond and Hausman might object to this example, arguing that what they propose would allow us to restrict the range of causes among which we consider allocating our support to only those for which we have some sympathy, since even our consumer choices are typically restricted to only those things we might actually enjoy. This may well be a reasonable argument, but it wouldn't necessarily excuse us from the requirement to consider potentially bizarre causes like cockroach preservation. Whether or not we are psychologically predisposed to endorse a particular cause depends somewhat on the level of specificity of the "good" we are considering. An environmentalist who supports species preservation should, in the unlikely event cockroaches ever became endangered, have some inclination to "save" them from extinction if she is to be consistent in her sympathies—and she should thus, given what the reconstructed argument suggests, consolidate her contributions accordingly.

7. This is a fairly close paraphrase of Michael Otsuka's (199) characterization.

8. It would, of course, offer similar guidance in ranking *obligatory* actions under consequentialism, and this is largely Parfit's concern, as well as Murphy's. Given the arguments in Part I, moreover, it may not be obvious that the contributions in question here would be a matter of optional charitable donations rather than obligatory (and enforceable) contributions to public goods. For the sake of the current argument, assume (1) that the charitable contributions at issue are supererogatory rather than obligatory, (2) that the principle of rational beneficence with which we are concerned is limited to just this supererogatory scope, and (3) that the question of how the rational altruist should best allocate her beneficence is distinct from questions concerning the obligations of free riders and honest holdouts.

9. The scenario is a mere reworking of the Drops of Water case—the example more commonly employed to demonstrate the elusive nature of clear moral principles where groups of altruists are involved, and used extensively by Otsuka (1991), who derived it from Parfit (1984:76), who himself adapted it from the work J. C. B. Glover. The classic case describes one thousand altruists, each of whom have the options of contributing one cup of water to a large tank which will be rolled into a battlefield to

quench the thirsts of a thousand soldiers, or of contributing the full cup to a moderately thirsty person nearby. Each, like Millhouse in our example, thus has the opportunity to fully quench one moderately thirsty person, or contribute what will amount to, once the water is again doled out, only a few drops of water each to a number of very thirsty persons.

10. One might object here that, as sympathetic as Millhouse might be to the plight of the homeless man, if he really has 25 dollars earmarked for an environmental cause, feeding the poor would not be an adequate substitute; charitable contributions, one might argue, are not all perfect substitutes for each other. Even if this were true, we could merely exchange an individual environmental effort for the homeless man's lavish meal—say, hiring a local youth to clean up one's neighborhood (or foregoing 25 dollars in wages to stay home from work and do it oneself)—and preserve the conditions of the example as well as its implications for the principle of rational beneficence.

11. The example which follows largely adapts our Millhouse scenario to that used by Otsuka to describe the paradox on pages 140–141 of his article, and the conclusions drawn from it are consistent with Otsuka's.

12. Otsuka (1991) himself offers a variety of reasons for believing that the ostensible "paradox of group beneficence" is really no paradox at all, among them the fact that treating it as such would commit one to the belief that the whole of the benefits to be conferred is somehow greater than the sum of its parts, or

$$(B_n\text{-}B_{(n-1)}) + (B_{(n-1)}\text{-}B_{(n-2)}) + \ldots(B_1\text{-}B_0) < B_n$$

where B_n denotes the benefits secured by n contributors (see pages 135–136). Otsuka argues that this notion is indefensible.

Chapter 11

Values, Preferences, and Category Mistakes

If I were to ask you what an automobile is worth to you, you might, without prompting, express its value in monetary terms. If, on the other hand, I were to ask what "your freedom" is worth to you, your response would likely be quite different. If you were to provide any response at all, it would probably be a less precise one than the x dollars (or Euros, or baht, etc.) that you are prepared to pay for an automobile, and might well approach infinity. More likely, you might reject the question entirely, arguing that one "cannot put a price on freedom," and suggest that I don't understand what I'm asking when I raise the question. All well and good, I say, but you were able to price automobiles when I asked. Suppose we adapt the question to this fact: you apparently know what an automobile is worth to you in monetary terms—how many of those automobiles, then, would be worth your freedom? How many would you require, in other words, to give up your freedom in return?

As is widely noted, such lines of query are often met with similar "protest" responses in CVM surveys attempting to price environmental or other public goods. Why is this, though? Why is it that, when asked to price something like one's freedom or the air one breathes, respondents often indicate that they would spend anything to purchase it or that it would take an infinite amount of money or another good to "buy" such things from them, or that, in what is presumed to be the same type of response, there is *no* amount which they would put on things of this nature?

One might suggest that the question makes sense economically but that we have trouble stipulating a personal exchange rate of money or automobiles to freedom simply due to the inordinately disparate "units" of each that are involved: so many automobiles would be worth one unit of freedom to us that we see it, for all intents and purposes, as an infinite amount, or an amount so large that it's essentially inestimable and pointless to try and capture.

This would be exacerbated by diminishing marginal utility: as we acquire more and more units of automobiles, the utility we might gain from each is increasingly less than the last unit. This is not true of "freedom," one might argue, since it is not something that can be readily disaggregated into discrete consumption units, two of which I can enjoy today, one tomorrow, etc. Thus it's not quite clear that, even if the question made sense, the two amounts could ever meet to establish an exchange rate.

But, the economist might maintain, we can indeed disaggregate things like "freedom" in the abstract into myriad *kinds* of freedoms, and these may be traded against goods like automobiles. As odd or unpalatable as it might sound, it's not nonsensical to say that one is willing to give up freedom of affiliation for an automobile, but that one will not trade freedom of speech for three Toyota Camrys. If in fact one is not willing to trade against particular types of freedoms, or if there is a level or amount of freedom below which one will not go, regardless of what might be offered in exchange, one is said to have a *lexicographic* ordering of her preferences.[1] Lexicographic preference orderings are those that assume a hierarchical structure such that certain preferences have priority for the individual, and there exist minimal amounts of their satisfaction that the individual must have before she is willing to trade additional amounts at the margin.

The notion of lexicographical preferences might be more readily understood in contrast to conventional indifference curves, the graphical depiction of the state of indifference that I've referred to numerous times in previous chapters. Economists since Alfred Marshall have suggested that it is for such indifference—not, say, "happiness" or "satisfaction"—that we strive, and the graphical image of such a state is now a staple of introductory textbooks in economics. The standard indifference curve between two substitutable consumer goods, A and B situates one good on each axis. The "curve" indicates the frontier of substitution: it marks the unit increase in A one would require for a unit decrease in the amount of B that she has, and vice versa. Figure 11.1 suggests what such a curve might look like between two particular goods, bread and tortillas.

There are a few items of note regarding indifference curves such as this one. For one, the presumed curve is typically convex, suggesting that there is diminishing marginal utility in substituting either good for the other; in other words, for each additional unit of bread we give up for tortillas, we desire increasingly fewer units of tortillas to substitute for the lost unit of bread, and vice versa.[2] This *marginal rate of substitution* between the two goods need not be symmetrical, as might be otherwise implied by the roughly semicircular nature of the curve here and the approximate parity in moving from point B to C and from point C to B. Had we drawn the curve decidedly flatter, this would imply that the amount of bread we might trade for one

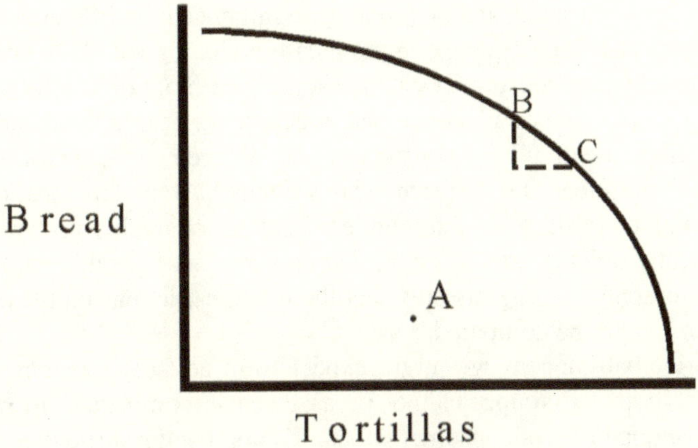

Figure 11.1 Standard Indifference Curve

unit of tortillas when we have very little bread and a lot of tortillas is different than the same trade we might make when we have very few tortillas and a good deal of bread. The same could be said, under these circumstances, for any segment along the curve, as well as under circumstances where the curve is steeper rather than flatter. Another point suggested by the curve in Figure 11.1 which should be stressed is that the curve itself represents the array of efficient trade-offs for an individual between the two goods. It depicts, in other words, the frontier of her relative desires where they are best satisfied. Unlike points inside the curve, we can say of any point on the curve that the agent is enjoying the maximum amount of bread she can enjoy given any particular amount of tortillas, and vice versa. Point A, for example, would mark an inefficient state; she could be enjoying more units of bread without a corresponding decrease in the number of tortillas she may enjoy, and, again, vice versa. This is why such curves are thought to be a heuristic portrayal of consumer indifference: at any point along the curve and only at these points, the consumer desires no more of either good that would not be offset by giving up a portion of the other.

Economists are, of course, well aware that consumers juggle more than desires for two goods at a time when making decisions about how they will satisfy their wants and achieve blissful "indifference," and attempts to model such states get decidedly more complex than simple two-good examples. The elementary case remains useful nonetheless, if only for purposes such as those at issue here. This is because lexicographic preferences represent a special case of preferences for goods that cannot be charted in this manner. When we pit things like freedom, basic sustenance, and perhaps environmental quality against other consumer goods, the claim goes, we encounter a threshold effect in which, below a particular level of such "goods," other consumer goods will not be traded, and no amount we might acquire of them will offset the insufficient level of the lexicographically valued good. This makes economic modeling of consumer behavior somewhat more difficult, but it does not, as far as economists are concerned, preclude the economic understanding of things like "freedom" as a consumer good, different amounts for which we might have preferences in the same way we may have preferences for different amounts of tortillas or automobiles. That lexicographic preferences are hierarchically ordered doesn't entail that they are incomparable within levels of their hierarchy; it suggests only that there are restrictions on the *degree* to which they might be compared.

The principal concern we might expect from advocates of environmental quality is that it simply cannot be assessed in comparison to monetary amounts or quantities of consumer goods. It's not that it is difficult to estimate what one would trade in dollars or Volvos to see significant improvements in the environment, or that such environmental goods are a "special case" of consumer goods for which we are compelled to make certain concessions in otherwise fine-tuned economic models, or simply that "pricing" the natural environment amounts to nothing more than a measurement problem. The point, as Mark Sagoff has argued in *The Economy of the Earth* (1988b) and elsewhere, is that we make a *category mistake* when we try to evaluate the natural environment in terms of what we are willing to *pay* for it:

> This is the kind of mistake you make when you predicate of one concept another that makes no sense in relation to it. If I say that the square root of two is blue, for example, I have not uttered a meaningful sentence but an absurdity, since color concepts are not of a logical type that can meaningfully be predicated of numbers. Likewise, if I tell you the window wants to be closed, you would not take me literally to attribute a psychological state to an inanimate thing, which would be an absurdity. Again, to ask for the name and address of the average American family is to ask a silly question. It is to be confused about the qualities that can meaningfully be attributed to an object and to commit, in that sense, a category mistake. (Sagoff, 1988b:93–94)

So too is the case, Sagoff contends, when we ask of one's public preference for one particular state of affairs in the world over another what he or she is willing to pay for it as if it were a private preference for mango sherbet over butter-pecan ice cream. The former is a matter of reason and doesn't submit to dollar amounts.

This quite illustrative passage from Sagoff does not itself lay out an argument on behalf of the claim that "pricing" the environment via subjective measures like CVM commits one to a category mistake, though it invokes, as does another classic work in this area, John O'Neill's *Ecology, Policy, and Politics* (1993), points that are expressed in argument (II) outlined in the opening passages of the previous chapter. In what remains of this chapter, I will be reconstructing systematically and sympathetically the various arguments for the categorical separation of environmental *values* and consumer *preferences* that are suggested in these works. I will maintain, however, that none of these arguments is adequate to establish that the environment cannot be "priced" and compared to other "goods" we might "purchase." I will argue in the next chapter that we have other grounds for doubting that WPT measures, be they established by CVM or otherwise, are capable of capturing the "value" of the natural environment. Thus despite the misgivings I will demonstrate here with the approaches of both Sagoff and O'Neil, we will eventually reach conclusions quite similar to theirs.

Let's begin by pinning down terms: just what is the *category mistake* at issue here? What is the difference between having a preference for particular automobiles or ice cream and embracing the notion of clean air or stable global temperatures? What is it that renders the latter a *value* rather than a mere consumer want? Consider one way we might lay out the argument:

1. If values were akin to preferences, they wouldn't conflict within the individual; that is, what one *values* would be what one *prefers.*
2. Individuals have a variety of values and preferences that are unique to the roles they play in society: consumer, professional, citizen, family member, etc. and the contexts in which such roles are called for.
3. These roles cannot be collapsed into a single, homogeneous role.

Therefore

4. There will be potential conflicts in values, preferences, and desires they hold across these respective roles which are irreconcilable.

Therefore

5. The values, preferences and desires one holds in these various roles cannot be assessed in the same way.

Let's call this the *argument from separate spheres*. We are, I take Sagoff to be arguing, mothers, physicians, voters, politicians, siblings, plumbers, etc., and we are, at times, consumers. There is no underlying role that unites the individual in all of these other roles, however many she may participate in, and the economist's attempt to reduce all of them to "consumer" appears no less arbitrary than reducing all roles a physician might inhabit in her life to "physician." Each of these roles orders its own desires and preferences, and we often find that these desires and preferences conflict in such a way as to prevent their meta-ordering from the perspective of an all-encompassing role:

> Attempts to find a "combined" or inclusive preference ordering . . . are bound to fail. They will fail for logical, not merely practical, reasons. Individuals have a variety of often incompatible preference schedules they reveal in the contexts appropriate to each, for example, in markets, family situations, professional contexts, and political circumstances. To try to combine these preference schedules into one is to search for a single comprehensive role the individual plays; it is to ask for the individual to behave not as a parent, citizen, consumer, or the like but in all and none of these roles at once. (Sagoff, 1988b:55)

In another passage, Sagoff offers a personal testimonial to the conflicts he himself experiences as a consumer and as a social and environmental activist:

> I love my car; I hate the bus. Yet I vote for candidates who promise to tax gasoline to pay for public transportation. I send my dues to the Sierra Club to protect areas in Alaska I shall never visit. And I support the work of the American League to Abolish Capital Punishment although, personally, I have nothing to gain one way or the other. (If I hang, I will hang myself.) And of course, I applaud the Endangered Species Act, although I have no earthly use for the Colorado squawfish or the Indiana bat. The political causes I support seem to have little or no basis in my interests as a consumer, because I take different points of view when I vote and when I shop. I have an "Ecology Now" sticker on a car that drips oil everywhere it's parked. (Sagoff, 1988b:53)

Despite the immediate appeal of Sagoff's argument, and of its apparent ability to jibe with many of our own experiences, it conceivably falls short on a number of grounds. For one thing, the inference from the premises to the conclusion (4) in the argument from separate spheres works only if we restrict our claim as to the potential conflict in desires and preferences to those that exist *across* roles. If we were to acknowledge that the desires and preference one holds *within* a particular role may well conflict, the argument no longer asserts anything interesting about the nature of these roles, at least not in a way that would tie particular types of desires to particular types of roles. A physician may value her autonomy as a professional and she may value the regard of her patients; yet in a system of private health care, most things she can do to satisfy the latter want will compromise the former: the flexibility of office hours depends on the demands of the HMO for whom she practices, the affordability of medications she prescribes depends on the profit margins of pharmaceutical companies who will dispense them, etc. It's not enough to say in such cases that her desires as a professional conflict with her desires as a consumer or as a political actor, or whatever (though there may well be conflicting preferences across such roles as well). The conflict seems to be between desires *within* her role as a physician. Somewhat more universally, we all encounter conflicting preferences as consumers and strictly as consumers; we want to spend as little as possible on an automobile, despite the fact that the add-on features appear tempting. If this is possible—if it may be the case that the things one values within a particular role can conflict with each other, and if we still believe it meaningful to speak of these roles and the comparability of the preferences or desires we have within them—there doesn't seem to be any need to abandon this comparability when we notice that our preferences and desires can conflict *across* these roles. In terms of the *argument from separate spheres* laid out above, we might suggest that the first premise is inadequate. If it is indeed reasonable to believe that one might have conflicting "values" as a professional or as a voter, and if it is reasonable to believe that she may have conflicting consumer "preferences," it is not unreasonable to believe that she may have "values" that conflict with her "preferences" in such a way that doesn't rule out the comparability of those that do not conflict with each other.

Another problem with Sagoff's particular concern is indicated by its reconstruction as the *argument from separate spheres* above. In order for one to accept the plausibility of premise (3), one must already assume the first inference to (4). In order to believe that the values we may hold in different roles are irreconcilable, in other words, one would already have to believe that there exists no dimension that underlies and orchestrates each of these roles, or that none of the roles are "more fundamental" than the others. But this is effectively what is being argued in the inference to (4) and subsequently to (5). The argument, in short, is somewhat circular, a fact that any economist with an unshaken faith in the claim that even the most complex of human actions can be reduced to utility maximization would surely point out. Much of both the acclaim and the disrepute that economists have achieved derives from their attempt to model all human institutions—even those we typically believe to be based on a variety of norms and elusive human emotions, most notoriously, perhaps, Gary Becker's (1981) analysis of marriage, family, and fertility. These models and the arguments on their behalf are by no means definitive, and other, quite brilliant economists have outlined the shortcomings of "rational actor" models, pointing out that their applicability is severely limited at best.[3] The availability of counterarguments notwithstanding, one cannot build an argument which presupposes the irreducibility of human values to calculations regarding pure economic preferences if this irreducibility and the integrity of these distinct values is precisely what one wishes to demonstrate.[4] It is precisely the ostensible conflicts between "values" and "preferences" that economists attribute to the irrationality of the belief that there is something to the former that can't be captured in the latter. If we are to counter the economist's assumption that values regarding the natural environment are merely preferences for environmental goods, this won't do it.

What Sagoff seems to want to argue in what I have reconstructed as the *argument from separate spheres* really involves the question of whether the variety of values, desires, and preferences we might hold can be reduced to a single measure: one's willingness to pay to see any such inclination fulfilled. This has more to do with the nature of how one might try to express a preference than with the question of the "roles" we play in expressing them. O'Neill (1993, 1997) more explicitly takes this approach in laying out a similar argument. We'll call this the *argument from incommensurability*; once again, I'll reconstruct it from the text itself, stripped down to its essentials:

1. For WTP measures to tell us what something is worth, at least weak commensurability must be true.
2. Weak commensurability is indefensible.

Therefore

3. WTP cannot tell us what something is worth.

In order to see how this argument works, we need, of course, to understand how O'Neill is using the phrase "weak incommensurability." To do this, we need to also understand what he refers to as "strong commensurability" and "weak *compar*ability."

Suppose you inhabited a very simple world with two other people, Patty and Selma, and only six objects in it (save for the land, sun, and sky that surround you): a deck of playing cards, a 9-foot square sheet of cellophane, a 4-foot tall propane tank, a goat, a book of matches, and a refrigerator shipping box. You reason that, in such a limited world, all of the objects serve best to shield the three of you from the relentless sunlight. Indeed, you assess each in terms of the number of ultra-violet rays it is able to block per square foot, and stamp this amount on the underside of each object for quick reference. Having done this, you explain to Patty and Selma, you have determined a measure by which all of the objects may be evaluated, and their value to the three of you can be ordered according to this measure, so that you can more readily decide which objects to preserve and protect and which ones are expendable. You point out that one can even make evaluations of the sort "the refrigerator box is twice as valuable as the pack of playing cards, and fifteen times more valuable than the sheet of cellophane"; in the event that more of the objects become available, you'll thus know how many would be necessary to substitute for each other.

Suppose that Patty agrees with your assessment that shade from the sun is the most valuable attribute that the objects have to offer. She doubts, however, that a single cardinal measure of "UV rays blocked per square foot" would be useful in assessing the exact value of each object. While she agrees that the objects can indeed be ranked in terms of their ability to provide shade, she dismisses your calculations as fatuous and simply ranks the objects in terms of their shade-providing utility, a ranking you note that does not deviate from yours: refrigerator box, propane tank, playing cards, goat, matches, and cellophane.

Selma agrees that the objects can be assessed and compared in regard to their value to the three of you, but contends that there is no single measure such as "shade providing" on which we can rank all objects. She notes that each of the objects has a different value in its own right. In addition to whatever sun-blocking qualities it might have, for example, the goat is a source of milk and thus necessary nourishment. Similarly, the propane tank, while shade during the day, can be used in conjunction with the matches to provide heat when the sun goes down. Even the playing cards have value as a source of leisure to pass what would surely be long, tedious hours in such a barren world. What each object is "worth" to you all, Selma thus argues, depends on the particular properties you might value in them. If you're hungry, the goat will surely be more valuable than the cardboard box; if you're bored, you would likely prefer playing cards to a propane tank. Hence each object can be "more valuable" or "preferable" to the others depending on what values are important to each of you at various times and under various circumstances.

If I am understanding correctly O'Neill's account, three distinct positions are being advocated in this contrived, limited world. You are asserting what O'Neill calls *strong commensurability*, the belief that there exists a "single property all objects possess which is the source of their value, and that our evaluative measure indicates the amount or degree to which that property is present" (O'Neill, 1993:103). Strong commensurability presupposes value monism; while we may think we value things on a plurality of different dimensions (and that their utility to us is no doubt among that plurality), strong commensurability suggests that this is illusory, and that there is really only one value to which all other values answer. Because, we assume, we can determine the precise amount of this property in all objects and states of affairs, we can make claims that presuppose an underlying, linear scale with a definite zero point—hence propositions of the sort "the refrigerator box is twice as valuable as a deck of playing cards."

Patty, on the other hand, is asserting only a *weak commensurability*. Like its counterpart in strong commensurability, weak commensurability holds that all objects and circumstances in the world may be ordered on a single, one-dimensional scale according to their value, and that this value is traceable to a single property common to each of them. Unlike strong commensurability, however, weak commensurability makes no assumptions as to the cardinality of that measure or the property it represents. Thus while, the advocate of weak commensurability acknowledges, we may rank all objects and states of affairs in the world according to the presence of this singular valued property, it is not a measurable amount that allows us to say some things are twice as valuable as others, while these things here are only one-third as valuable as those things. O'Neill suggests that weak commensurability presupposes *strong comparability*—the belief that all things may be ordered (and compared) in value, though he doesn't offer much of an explanation as to how it might be distinguished from weak commensurability (see O'Neill, 1993:103–104). For our purposes, we'll treat them as interchangeable terms.

Lastly, and in contradistinction to weak commensurability/strong comparability, Selma is advocating *weak comparability* in our example. Like its counterparts, weak comparability acknowledges that things in the world may be compared to one another, and that we may have rational preferences for some things over the others. Unlike its counterparts, however, weak comparability does not assume that there exists an underlying property which can order all objects according to a single value, let alone indicate how much of that particular property it has. It suggests that, at best, there is a plurality of values we use to evaluate objects and states of affairs in the world, and no single value presides over the others. While evaluations of things are undeniably rational, they are performed on a piecemeal basis, where one object or state of affairs is compared to another, and that one to yet another, based on whatever values are deemed critical to the particular comparison. Perhaps the most concise way to distinguish weak comparability from weak and strong commensurability is to note that, unlike the latter positions, weak comparability abandons the hope of transitivity between all objects. If we assume that an underlying property imparts value to all objects, we assume that if object A is more valuable than object B, and B is more valuable than C, we can infer that object A is more valuable than C, even though we ourselves might not have made a direct comparison. Weak comparability, in denying a single underlying source of value, will not allow such inferences, since there may well be other values which would make C preferable to A.

O'Neill's strategy in what I've called here the *argument from incommensurability* seems fairly straightforward. He would like to portray commensurability (strong and weak) and weak comparability as mutually incompatible—something which seems quite agreeable. Having done that, he then needs to argue that WTP measures and the methods which employ them[5] presume at least weak and possibly strong commensurability. This, O'Neill describes, is suggested by their roots in classical Benthamian utilitarianism, in which the hedonistic desire for pleasure and the aversion to pain are understood to be the cardinal sources of value. Even when we consider the modifications suggested by preference utilitarians, who would find value solely in the satisfaction of preferences (regardless of whatever pleasure this might bring the agent) and thus would measure individuals' willingness to pay to see those preferences satisfied, we are still describing what amounts to a presumption of strong commensurability (O'Neill, 1993:110). O'Neill more modestly stakes the argument on WTP measures' commitment to weak commensurability, however, for unspecified reasons.

Having thus defined commensurability and strong comparability as incompatible with weak comparability, and having argued that WTP measures are committed to at least strong comparability (and possibly commensurability), all that remains is for O'Neill to offer a defense for (at best) weak comparability. This he does by asserting the value-pluralism he takes to be concomitant with weak comparability:

> Evaluation of objects under different descriptions invokes not just different practices and perspectives, but also different criteria and standards for evaluation associated with these. It presupposes value-pluralism. Appeal to different standards often results in conflicting appraisals of an object: . . . an object can have considerable worth as an A, B and C, but little as a D, E and F. Given such value-pluralism, can one expect commensurability in the form of strong comparability, or must we be content with weak comparability? The answer depends in part on whether one is comparing objects within a perspective or across different perspectives. (O'Neill, 1993:108)

O'Neill believes that we indeed must settle for, at best, weak comparability, because our evaluations of different objects and states of affairs must compare across different perspectives and work from among different values. A stretch of wilderness, for example, may be valued for aesthetic reasons, for reasons of utility, or on the grounds of its intrinsic qualities. To "price" such a stretch of wilderness—to effectively abstract from these values and assign it an economic value at which it can trade against other goods—ignores this plurality of ways in which we might value it. Even if we forsake cardinal measures of value such as WTP and consider only our preference rankings, O'Neill argues, we can be left in a situation in which value-pluralism prevents us from developing a comprehensive ranking of the things we might value. He employs a curious device for demonstrating this: an adaptation of Kenneth Arrow's classic "voter's paradox." Where Arrow, however, described a hypothetical case of different voters with different options who, by design, cannot come to a democratically preferred outcome, O'Neill posits a case of different values held by the same individual—one who would order his preference for various outcomes differently and in such a way that no option would be dominant across values (O'Neill, 1993:111–112). O'Neill doesn't offer a concrete example of how this might work for particular values and particular options, but it shouldn't be too difficult to apply it to specific environmental goods. Table 11.1 represents, in the columns, three "values" to which one may adhere—that is, three potential aspects of the natural world which one might hold in high regard and for which, given tangible opportunities, one might be willing to pay: wilderness integrity, animal rights, and stable global temperatures. The rows of Table 11.1 represent three such opportunities—three options across and in support of which one might allocate one's funds: wetlands restoration, wolf repopulation to a lost habitat, and various offsets to carbon emissions intent on minimizing the adverse effects of climate change.

It probably wouldn't be all that striking to note that how we might rank the three environmental amenities in question varies with the values in question—typically in ways that would make intuitive sense, given a particular value. If, for example, we consider only wilderness integrity as a value, wetlands restoration might be our first preference, with wolf repopulation close behind, and the institution carbon offsets ranking third. Similarly, if we focus only on stable global temperatures, the benefits of carbon offsets might be our first choice, followed by wetlands restoration and wolf repopulation. What is more interesting about what Table 11.1 portrays is the fact that, when all three values are considered in tandem, there is no option which appears to be clearly more preferred than the others. Wetlands restoration is preferable to wolf repopulation if we consider only wilderness integrity or global temperatures, but this is not so when we consider animal rights as well. By the same token, wolf repopulation is preferable to carbon offsets if wilderness integrity and animal rights are the only values we consider in ranking our preferences, but this dominance disappears once we consider global temperatures. The same can be said of every comparison we might consider in the table: there emerges no dominant option which we could say is preferable to at least one other option under all values. O'Neill concludes from comparisons of this sort that the value-pluralism which is endemic to all of our normative decisions would preclude the grand ordering of preferences along a single dimension—certainly one consisting in dollar amounts or bundles of other goods—which he presumes to be entailed in WTP assessments of economic "value."

The upshot here is that, unlike Sagoff's argument, which suggested "values" and "preferences" belong to separate spheres of our lives (the political

Table 11.1 Ranking of Environmental Preferences and Options

	Wilderness Integrity	*Animal Rights*	*Stable Temperatures*
Wetlands Restored	1st	3rd	2nd
Wolf Repopulation	2nd	1st	3rd
Carbon Offsets	3rd	2nd	1st

and the economic), O'Neill seems to be acknowledging that preferences and values work together in some meaningful way. What runs contra to the assumptions of CVM or any other method which assesses value strictly as one's subjective willingness to pay for a good or a state of affairs, O'Neill suggests, is that these preferences are themselves ordered by our values—they are an *effect* of values rather than their source. Since these values are numerous and varied, and since each may potentially order our preferences in different ways, it is pointless to hold out for a seamless ranking of all of our preferences into a single lengthy index:

> Even if pleasure were the ultimate and only intrinsic value, it could not provide a single value to order goods, since pleasures themselves are plural in character: the pleasure of drinking beer and that of good conversation are different in kind and are not measurable on a single scale. Preferences, on the other hand, answer to values—not values to preferences. I prefer A because of its value, I do not value it because it is preferred. Hence, given plurality of values, our preferences merely record our judgements in resolving conflicts between them. They do not provide the supreme value through which they are to be resolved. (O'Neill, 1993:110)

Economists, according to O'Neill's argument, have simply misinterpreted economic behavior to be an end in itself, much like we might conclude that a man stooping to pet a small animal desires to be shorter. When they encounter preferences for things which are informed by a variety of values, and when those values do not make themselves evident, they are left to work with the only tools they have—tools which have a limited range of application at best.

Does O'Neill's argument give us firmer ground than did Sagoff's in asserting that CVM is committed to a category mistake in conflating values and preferences, and that WTP measures are thus an inappropriate indicator of the value of environmental amenities? Possibly not. It is by no means clear that welfare economists prone to endorsing WTP measures as surrogates for economic value are required to accept strong or even weak commensurability, though the first premise in what I have called here the *argument from incommensurability* certainly wants to suggest that WTP measures favored by some economists entail that at least weak commensurability is true. O'Neill's own case for this claim, as we have seen, rests largely on the historical debt welfare economics owes to utilitarianism—both in its classical, hedonistic forms exalting *pleasure* (and the avoidance of pain) as the primordial good and in its revised forms which deem *the satisfaction of preferences* the ultimate aim toward which humans strive. Such a debt is arguably overstated. Since the rise of ordinalist welfare economics in the early twentieth century, observed economic behavior has been thought to provide information strictly about the relative *priority* of one's preferences concerning the goods available to her, providing little if any verification of cardinal utilitarian assumptions about *what* it is that is being satisfied. Despite seeming similarities, the ordinalist's ranking of preferences is quite different from even *preference utilitarianism*. Utilitarianism in any of its various forms carries with it a theory of what is ultimately *good* and a theory of what constitutes *right action* (i.e., how best to achieve the stipulated good). Thus preference utilitarianism asserts no less than classical utilitarianism that utility is good and that people ought to act in such a way as to maximize it. In the case of preference utilitarianism, this utility is found in the satisfaction of preferences (rather than, say, the provision of things we might believe would bring one pleasure, whether one consciously prefers them or not), and we are therefore obliged to take actions which would best maximize people's preference satisfaction. Economists since the emergence of ordinalism, on the other hand, as it is often lamented, are neutral as to a theory of the good. When they assess individual preferences and advocate policies which, they believe, will best satisfy these preferences, it is not with the ultimate purpose of "maximizing preference satisfaction" but rather with the purpose of enabling individuals to best achieve their particular ends, whatever those ends might be.[6] Hence the trappings of utilitarianism—namely the assumption that there is something intrinsically good in any of the ends economics can facilitate—are all but abandoned in most of the economists O'Neill must surely have in mind. The issue was one of far-reaching debate among economists a century ago, and it's fair to say in retrospect that the ordinalists triumphed in the disputes, signaling the demise of marginalism and the weighing of cardinal utilities, if not the very marriage of moral philosophy and economics that had been referred to as *political economy* for over a century. Vilfredo Pareto was a leading advocate of this transition, and the general position is summarized succinctly by one of his successors, John Hicks. As utilitarian, marginalist economists such as Alfred Marshall and William Stanley Jevons understood it, Hicks explains,

> 'given wants' is interpreted as meaning a given utility function, a given intensity of desire for any particular collection of goods. This assumption has made many people uncomfortable, and it appears from Pareto's work that it is not a necessary assumption at all. 'Given wants' can be quite adequately defined as a given scale of preferences; we need only suppose that the consumer has a preference for one collection of goods rather than another, not that there is ever any sense in saying that he desires the one collection 5 per cent more than the other, or anything like that. . . . If one is a utilitarian in philosophy, one has a perfect right to be a utilitarian in one's economics. But if one is not (and few people are utilitarians nowadays), one also has the right to an economics free of utilitarian assumptions. (Hicks, 1939:18)[7]

Still, one might object, even if the economist is committed only to assessing and enabling the satisfaction of personal preferences and not "maximizing preference satisfaction" (or any other ultimate "good"), doesn't this still suggest weak commensurability is true—that individuals must be able to order all of their preferences on a single scale? This question is not one to which many economists speak directly, but their professed value neutrality—whether or not it is defensible—does suggest that it really shouldn't matter for purposes of their value-assessing tools whether or not individuals are able to order their preferences on a grand scale. Whether I prefer wetlands to wolves to carbon offsets (and even the relative amounts I might be willing to pay for each) depends upon how I order my values. Similarly, we don't really put a "price" on "freedom"; if, however, "freedom" is among our values, we might weigh certain things (e.g., certain types of newspapers, the ability to gather with politically like-minded individuals) over others and, yes, we might be prepared to estimate how much these things are worth to us in terms of other things. *If* I can order the values that inform these preferences from those to which I am the most committed to those which I am the least committed, I might well be able to order all of my preferences as well. If, for example, I discover that wilderness preservation is more important to me than animal rights or global temperatures, I can be fairly confident that I prefer wetlands restoration to wolf repopulation to planting trees to collect carbon emissions. Nothing about economic theories adhering to WTP measures as assessments of economic value seems to require this ordering of values, however. Ideally, the "rational agent" of modern economic theory might have Plato's well-ordered soul, but it's by no means clear that such an individual is required for economic price theories to "work"—at least insofar as they can accurately reveal what individuals are willing to pay for goods and states of affairs. Indeed, the economist might argue that since we need to "discover" in some sense which of our values take priority over others, assessing our preferences, and even putting relative economic prices upon them, is one efficient way of revealing which values really are most important to us.

Hence WTP measures don't seem to require that strong or even weak commensurability of all preferences, desires, values, etc. be true. Rather, they appear to be quite indifferent to the relative commensurability of these things. CVM or any method of economic pricing doesn't really care about our ends in preferring A to B, or about the quality of our *reasons* for assessing A and B at particular relative values, or even if we *have* reasons for such evaluations—moral or otherwise. It takes our preferences, for better or for worse, at face value, and cares little whether or not they may be justifiable.

This very point, ironically, could well be the Achilles' heel of CVM and WTP assessments of value, and suggests what may be yet a third argument on behalf of the claim that such assessments are committed to a category mistake.

An example may help to illustrate it before formally laying it out. A substantial portion of the book you are reading began its life as my Ph.D.

dissertation in philosophy. Imagine that I wrote it with the expectation that it would not be evaluated in the manner doctoral theses are normally evaluated. Imagine that it was not to be submitted to a dissertation committee for the members' review, edited and revised according to their comments, defended before them as well as an audience of peers, and ultimately brought before my university's graduate college with a petition from all those involved stating that I have mastered the tools of my profession sufficiently to carry the title "Ph.D." Suppose instead that we intended to survey other philosophers, environmental activists, economists, and similarly interested parties to determine what each would be willing to pay to see my dissertation accepted and a degree conferred. If this amount were to surpass the aggregate amount elicited from still other interested parties who would be willing to pay to see this thesis *not* be given a passing grade, I would be granted a doctorate in philosophy. If not—if others are willing to pay more to see the thesis shunned and ignored—no degree would be conferred.

Fortunately, this is not how my dissertation was evaluated, nor how any dissertation in any field is likely to be evaluated in the foreseeable future (including, it's worth noting, dissertations in economics from the University of Chicago or other bastions of free-market thought). Why is this? We would probably attribute this tendency to the fact that we do not see the proper evaluation process for an academic work as one of determining the relative strength of the aggregate preference for its findings, arguments, etc. We judge academic work, we like to believe, on its merits alone, and not on its ability to elicit substantial WTP amounts. While it is certainly conceivable that an academic thesis may have commercial appeal, we would think this to be a contingent fact, and not something that is essential to our evaluation of its quality (since it is equally conceivable and, regrettably, more likely for a top-notch doctoral dissertation revised into a monograph format to end up collecting dust in Amazon.com warehouses). Either way, we generally believe that whether an academic work worthy of a degree is something that is subject to reasoned debate (is it sufficiently original?; does it further knowledge in its area?; does it properly and sufficiently acknowledge previous work in its area?; etc.) and not a matter of sheer "preference."

This line of argument is picked up by both Sagoff (1988b:44–45, 54, 113) and O'Neill (1993:68, 76–81), each of which point to a difference between those things that are available to scientific scrutiny, normative evaluation, and other standards, and those which are subject to simple tastes and preferences. Only the former, the argument suggest, require *reasons* for our evaluations, and to treat them as preferences requiring no justification would be to commit the category mistake under examination here. We'll call this one the *argument from justification*, and reconstruct it in the following manner:

1. *Preferences* can be assessed on the basis of the intensity with which we are willing to pay to see them satisfied.
2. *Values* concerning how the world ought to be require normative justification to be considered acceptable.

Therefore

3. Whether or not we hold a particular value is independent of our preferences.
4. Public policies concerning the things we value ought to be aggregated from those values.

Therefore

5. Public policies should not be determined by pure preferences.

A preference for Thai food is just that—a preference. One might offer an *explanation* as to why one prefers Thai cuisine over, say, French, but a justification based on the relative merits of Thai food would not be required for one to be willing to pay to consume it. All that matters is that one has a preference for it; we can accept this at face value, and we might even let markets do their work and allow that aggregate preference to (at least partially) determine the availability price of Thai food. A "preference" for wilderness preservation, the argument suggests, is a different matter. Whether or not it is "worth it" to preserve wilderness areas is a function not of how much people are willing to pay to see them preserved, but rather of the quality of the arguments on behalf of the policy. We evaluate, in other words, not the aggregate demand for wilderness preservation but rather the *reasons* for it. This is why environmental policy is different from a market, the argument suggests, and why a "preference" for stable global temperatures is different from a preference for a soup made from coconut milk, basil, and lemongrass. The former requires that we supply reasons why the good should be provided and an adequate argument that it ought to be provided is sufficient reason for providing it, regardless of what one is willing to pay to see it provided; the latter requires about $8.95, and will be provided to the extent we are willing to pay for it in a good Thai restaurant.[8]

As Sagoff and O'Neill both acknowledge, this difference is more readily apparent for some goods than for others. Certain things are clearly in the realm of those which require scientific, historical, political or moral justification and, as in the case of my dissertation, we would rightly think it odd if their justification abandoned appropriate standards of evaluation and were replaced by the willingness to pay of relevant interests:

> A market or quasi-market approach to arithmetic, for example, is plainly inadequate. No matter how much people are willing to pay, three will never be the square root of six. Similarly, segregation is a national curse, and if we are willing to pay for it, that does not make it better but only makes us worse . . . Similarly, the war in Vietnam was a moral debacle, and this can be determined without shadow pricing the willingness to pay of those who demonstrated against it. Similarly, we do not decide to execute murderers by asking how much bleeding hearts are willing to pay to see a person pardoned and how much hard hearts are willing to pay to see him hanged. (Sagoff, 1988b:44–45)

We of course, Sagoff concedes, make bad decisions or offer inadequate answers in each of these instances. But these would be failures in the relevant mode of discourse: arithmetic, moral philosophy, scientific hypothesis testing, etc. They wouldn't be "market failures," since there are no markets here within which to "fail."

Given that the inappropriateness of WTP measures seems obvious in the case of mathematical truths, scientific theories, and political/moral debates, Sagoff argues by analogy that they should also seem inappropriate in the establishment of environmental policy. We judge wilderness preservation, the protection of wildlife habitat, the containment of greenhouse gas emissions, etc., to be valuable on the basis of what we believe is right, and it makes no more sense to ask what these things are worth to us than it does to ask what it might be worth to us that capital punishment be "right" or "wrong," or that 24 be a prime number, or that gravity be revoked.

Despite the strong intuitive appeal of the *argument from justification* and of the arguments by analogy both Sagoff and O'Neill employ to bolster the third premise, I think the argument fares no better than the first two. Individuals *can* have preferences about the state of the world that appear either consistent with or contrary to what they know about it or what they judge to be true—be those truths mathematical, empirical, or moral. Granted, it would be silly for one to acknowledge that 24 is not a prime number and be further willing to pay a substantial sum to "see to it that it isn't" (or, more curiously, to "see to it that it is") by affecting elementary school mathematics textbooks. It would be decidedly less silly for one to believe that rape victims should have reasonable access to abortions, or that one nation should not carry out unprovoked military adventures in a sovereign nation, or that capital punishment is barbaric—and for one to be willing to pay $x to see that public policy reflect these values. These beliefs and preferences are not inconsistent, and it seems at least plausible that one might derive somehow an amount which represents the importance of these values to oneself. We are not in the realm of the inconceivable.

The problem, more precisely, seems to be that premise (3) is referring to the intensity with which we might desire that a particular normative *claim* be *true* (put aside empirical and mathematical claims for now), while the conclusion (5) is referring to the intensity of *preferences* we might have for particular *policies*. Sagoff and O'Neill are attempting to defend moral realism (and realism in general) with the argument, suggesting that how badly we might want a normative proposition—for example, wilderness should be preserved—to be true is irrelevant to the truth of the proposition; this is essentially what premise (3) proposes. Moral realism, however, is not at issue in the conclusion. It is possible for one to accept moral realism, to accept that normative proposition P is justifiably true, *and* for him to prefer that public policy be directed independent of even in spite of this fact. The *argument from justification*, as reconstructed above, thus unduly conflates these two things—the "preference" that a moral claim be acceptable (agreeably, a nonsensical preference) on the one hand, and the desire that, regardless of its moral standing, a particular public policy be implemented on the other.

Sagoff, O'Neill, or any advocate of the *argument from justification* as I've reconstructed it here may well raise the following objection at this point: "This is not how we typically conceive of moral values—as things which we can believe ought to be true of the world but for which we must *pay* to see enacted as policy. If we hold a particular normative value such as 'wilderness ought to be preserved', we hold it under the assumption that it is something that should be believed not just by ourselves but by everyone. Among the things that makes a particular normative judgement something more than a preference is precisely the fact that we can coherently believe that everyone else should also have the same normative belief; in Kantian terms, it is a moral value because it can rationally be *universalized*. It would thus be nonsensical to believe that 'wilderness ought to be preserved' but to also believe that what we actually *do* with a wilderness area should be left to the caprice of what individuals are willing to pay to see it preserved or, worse, what wealthy corporations can pay to see that it isn't."

If one wishes to argue this, one can, but this is not the argument from justification as I believe Sagoff and O'Neill intend it. The "values" to which the argument from justification refers concern beliefs we have about states of affairs in the world themselves. The "values" this objection would attribute to the argument from justification, on the other hand, refer to beliefs we might hold regarding what people *ought to believe* about states of affairs in the world—namely, that they should not believe the value of the natural environment is open to the sway of the economist's WTP criterion. Whether or not such judgments may be universalized is not at issue here, since universalizability may be characteristic of either sort of belief. That wilderness is desirable may well be true, and it may well only make sense as a normative judgment if it is something we must believe true for all rational agents. It is a different matter to believe that *others ought to believe* that wilderness ought to be allocated by something other than markets. People may have preferences for enjoying environmental goods and they may have preferences that people conceive of environmental goods as something other than preferences. These may both be considered values, but there are decidedly different things we can say about these values. My desire that wilderness be preserved may well be a belief that is better subject to normative justification than to an appeal to my "tastes," but my desire that the question of whether wilderness is preserved or not be something that is decided not by markets but by a political process—perhaps a desire subject no less to normative justification—would constitute a different value. The argument from justification appears to equivocate on these two uses of "value"; it wants to draw conclusions about the latter but only speaks to the former. This isn't to suggest that either or both are inappropriate uses of "value," or to deny that environmental goods fall more properly into either sense of "value" than into the category of "preferences," but it would require a different argument from the argument from justification to conclude that public policies should be the culmination of the values we hold and not our preferences.[9]

I think that if we were to make such an argument, we would find that the moral issue here hinges not on the difference between "values" and "preferences" (or the relative commensurability of these, or the respective "spheres" in which we hold them) but rather in the differences among what our inclinations (call them "values" or "preferences") *require of others*. Consider the issue in this way. People have desires for certain things and for particular states of affairs. The nature of those desires and of those states of affairs will undoubtedly vary with the circumstances under which individuals live, but it remains more or less true that people have them. Some of these wants may be fulfilled by the efforts of the individual alone. Under certain circumstances, this may mean one's own efforts are all that is required to fulfill the want (put aside, for the sake of argument, all of the historical efforts of others that brought the individual to this situation), under others it may require one's own money obtained via her efforts and under the conditions of an economic system. Other wants, by their nature, require the cooperative efforts of any number of individuals, occasionally many. Again, how this cooperation would be articulated and orchestrated depends on the nature of the political/economic system under which they live, but there will, inevitably, be such wants. These things entail costs no less than those obtainable through individual efforts—often greater costs—and are generally not things that an individual can provide for herself. The difference between the two types of "goods" that may be desired is not that the former merely satisfy economic "preferences" while the later constitute *values* about the way we think things *ought* to be. We can quite coherently believe that conditions *ought* to be such that they facilitate us in satisfying our personal appetites as readily as we believe things ought to be such that air remains unpolluted or that wilderness areas we will never see are preserved. We may, moreover, be as prepared to pay to see that the latter desires are satisfied as we are the former. The difference is we must also believe of the latter that *others* ought similarly to pay something to see such desires satisfied. The preference for an environmental amenity is thus, in important ways, a preference no less than a preference for curry over pasta. What marks it as something of special, political concern is that it is a preference we wish others to hold and be willing to contribute to as well (while it really needn't matter to us if everyone else prefers pasta to curry). There is undeniable normative content in the nature of the preference itself as well as the question of whether others ought to hold it; again, there is no dispute here that it is possible to normatively defend claims such as "wilderness ought to be preserved" and that believing the claim should entail the belief that it is something others ought to believe as well (and that such beliefs are not required when preferring curry to pasta). The normative content that has greater implications for how we assess the value of the natural environment when making social policy, however, lies in the fact that we believe that others, if they too hold such a preference, *should also bear some of the cost in seeing it fulfilled* (though, again, what this means in a particular case will depend on the economic circumstances within which one holds the preference). We can ask of an individual what she might be willing to pay for either type of good—curry or clean air. She can, in turn, provide an answer to either question. What is different about her WTP estimate for clean air is that it implies, whether she realizes it or not, the claim "other people are breathing it, and they should be paying their fair share as well."

What does this imply substantively here? For one thing, it implies that economists who argue that values are not "free" and who advise that we "price" the environment in making social policy have a more or less reasonable claim. While this claim alone falls short of articulating the array of normative issues at stake, it suggests that so too will the claims of environmentalists for whom the moral questions raised in the provision of environmental goods have more to do with "intrinsic value" and the "rights of nature" and less to do with equity and fairness. Providing a healthy, viable environment entails costs, whether we choose to acknowledge them or not, if only the opportunity costs of production routes foregone. My demands that a stretch of wilderness be preserved as such so that I may travel to it and camp in it at will (or simply that it be preserved for the sake of its existence) would preclude a number of other fates for the same stretch of land that others may have in mind. Neither the fact that I have the superior justification for my proposed fate (keep it wilderness) nor the fact that a majority of the population may share my moral judgment would, by themselves, be sufficient grounds for a policy which diverts land use toward these ends. Such policies must also consider whether I am willing to bear the costs of preserving wilderness and, since I could not be expected to bear it myself, what a fair distribution of that burden might be. Similarly, as we saw in the Introduction, my demands for social investment in environmental quality will entail that wealth not be spent on other, arguably noble goals, such as the amelioration of poverty or the international enforcement of human rights. For me to thus say "I value the natural environment in such a way that I can't put a price on it" offers little consolation to those whose immediate interests sit with these other causes. What is morally significant about our "values" which entail the provision of public goods is that others will have to contribute to see them enacted. This doesn't deny—I cannot stress this enough—that values can and ought to be defended rationally, and that we can coherently argue that others ought to hold them. There are, in most cases, good reasons why a national park ought not be razed and a theme park built on top of it, and there are good reasons for believing Disney's Co.'s willingness to pay billions for the opportunity to do the razing does not trump the first set of reasons. The point here is that our reasons for preserving the land as a park—good as they may be—do not displace the fact that the values we hold for the land (and that we, by the very nature of the good, wish others to hold for it) must be backed by willingness to bear the costs of seeing the preference fulfilled. If the values we hold are for things or states of affairs that will require that we ourselves and others shoulder some expense, we need to consider how much of this expense we are willing to bear. Granted, how we arrive at such a value is by no means clear. But it's not entirely unreasonable that we might think, at some basic, pre-theoretical level, that we can get at this value simply by asking people "What is the natural environment (or this particular amenity) worth to you?" If this is inadequate—if we are to forego the purely subjective assessment of how much of this burden we are willing to bear for other types of assessments—we would need good reasons for this. I don't think the types of arguments considered thus far in Part II provide such reasons, though I do believe that a case can be made.

By this point, it appears that we can safely if tentatively conclude that, if they are to be meaningful at all, CVM surveys or any similar method of eliciting what individuals might be willing to pay to ensure the provision of environmental goods (at least those that approximate public goods) would need to be explicitly designed as a sort of tax referendum—that is the question "What would you be willing to pay for environmental amenity X?" would need to be framed in such a way as to indicate (a) this is not something the individual is going to be able to provide herself and (b) the good will be provided only if enough people estimate their willingness to contribute toward its provision at a level sufficient to cover its costs. In some cases, the taxes in question would be intended less as a revenue source to provide an environmental good than a disincentive for the activities known to impede the natural production of the good or hasten its depletion; in such cases the question might reasonably be translated as "What sorts of lifestyle changes are you willing to undertake to provide or preserve this environmental good?" Even this, however, leaves open the critical normative question of *how much of this burden we believe others ought to be willing to bear.* This is, of course, a non-issue with goods that can be allocated on private markets; while we believe that others should pay a similar price (why should we alone be gouged?), we accept that this price is a function of our aggregated WTP values estimated without regard to the obligations to others. Charitable contributions, moreover, as I argued in the last chapter, do not carry such implications. When we donate our resources toward a cause, we may do so rationally in the hope that our contribution makes a difference, but we do not estimate the "proper" contribution on the basis of what we believe others will contribute, nor necessarily on the basis of what we believe they *ought* to contribute. Hence, as we saw, the concern that WTP assessments were actually "charity" in disguise were misplaced. I am now suggesting that the real problem here is this: the nature of the goods in question entail that if they are to be provided at all, they will require "donations" from others. But such "donations" cannot be supererogatory; if we believe the good is something that ought to be provided, and if we are willing to shoulder some of the burden in providing it, we must also believe that others must shoulder some of that burden. And if this is the case, the amount which we might be willing to pay ourselves can only be conceived as a proportion of the amount that will be necessary to provide the good. This clearly doesn't seem to be what we are eliciting even when we ask "what would you be willing to pay in increased taxes to see the government provide environmental amenity X?"—a fact not entirely lost on the economists who actually work with such methods, at least one of which is prepared to concede this as a shortcoming of CVM and of WTP measures in general when solicited in regard to public goods. Daniel Kahneman himself—whose considerations on the *embedding effect* we examined in Chapter 9, notes quite eloquently that

> it is likely that the value of a particular product of social action to an individual depends strongly on the details of how that action is performed—for example on the equity of the distribution of payments.
>
> There is a bind here: we intend the CVM to mimic what a free market would generate. But a free market is inconceivable for many of the goods that we wish to value. The only realistic way to achieve some goods is by government intervention or by social action, and the cost of this action must be distributed, either progressively or equally, among members of the community. In such cases, it is indeed impossible to separate the value of a good from the procedure by which that good is obtained. In particular, WTP will then depend on others'

payments. Note that this is a concern for equity, which is not the same as a strategic attempt by individuals to minimize their payments and maximize their benefits. What happens here is simply that if I am asked to pay $50 to preserve Ontario fish, I would like to know who else is going to pay $50. (Cummings, et al., 1986:194–195)[10]

Sagoff and O'Neill are on the mark in suggesting that there is something normatively lacking in WTP assessments of environmental value. The problem, however, is not simply that questions concerning the provision of environmental goods cannot be answered with prices. Rather, the problem is that such questions cannot be answered without reference to the proper burdens we may place on others in providing them.

"Fine," we might conclude. "The relevant question is not simply 'What are you willing to pay for environmental good X?' but rather 'What are you willing to contribute toward the provision of environmental good X given that it is a burden that must be shared among the population who will enjoy it (be that population two people or seven billion)?.' Let's ask it and get on with environmental policy." Advocates of CVM who might maintain such a position are in effect suggesting that this is a methodological question of accommodating the *strategic incentives* to inflate or deflate one's stated WTP in relation to one's "true" WTP. What one is willing to contribute toward a collectively provided environmental good is, no less than a strict WTP for a private market good, a "primitive" preference—one that can be revealed by economic activity or elicited by the right question and the proper "demand revelation" techniques, even to the point of forcing one to betray one's "strategic bidding." In the next chapter, I argue that the empirical concern with what one believes others are *likely* to be willing to pay for collective environmental goods (and the attendant adjustment of CVM tools) is distinct from what one believes others *ought* to pay. Normative beliefs about the proper burdens others should bear in providing public goods, I suggest, must be formulated before one can offer a coherent assessment of what one is "willing to pay" for such goods. If defensible, such a claim would not sit well with those who would ask us to believe that what we might pay toward the provision of an environmental good is pure economic "preference."

NOTES

1. See, for example, Gowdy and Olsen (1994) for an instructive account of such preferences.

2. Hicks (1939:14–16) points out that the fact indifference curves appear to reflect diminishing marginal utility is a contingent rather than a necessary one ("utility" is not something in which Hicks had a good deal of faith). Hicks' general discussion of

indifference curves in Chapter One of his *Value and Capital* is especially lucid and clarifies a number of the important subtleties I'm glossing over here for the sake of simplicity.

3. See Amartya Sen (1977) for one of the best examples of such critiques.

4. Curiously, though it doesn't undermine the argument itself, one might question the strategic value of Sagoff's defense of the separate spheres of human activity, and the irreducibility of the values we hold in each. One of the principal criticisms of the effects of capitalist culture—attributable largely to Marx but taken up in innumerable veins—points to its bifurcation of the human individual as "consumer" and "citizen," and the harmful effects this separation brings to him and to society at large (recall that I touched briefly on this issue in Chapter 7 in regard to classical Liberalism). One would think radical, anticapitalist environmentalists willing to endorse Sagoff's general concern about the status of environmental values would be hesitant about basing such a status on a human psychology that is arguably a by-product of capitalist economic relations themselves. Alan Holland (1995:31–33), acknowledging, as we have in the last chapter, that consumer preferences themselves are rarely of a pure economic nature, revises Sagoff's argument along Platonic lines, noting that preferences aimed at satisfying our appetites are not the same as the "second-order" preferences which may express attitudes *about* those appetites. Still, Holland's argument would be no less vulnerable to the problems noted of Sagoff's here.

5. O'Neill's explicit target in these discussions is Cost-Benefit Analysis rather than CVM per se, though he understands WTP measures to be the principal means of assessing costs and benefits. I had suggested in the Introduction that this is a contingent fact of CBA—something to which O'Neill might well assent.

6. The distinction here may seem strained, and it is admittedly a subtle one. Consider it in terms of a smaller, concrete example. Hoover, Skinner, and Chalmers each have preferences for particular things they believe would make them happy. Some of these preferences may overlap, some are unique to one individual or another. We can say of the preference utilitarian both that she believes there is good in the preferences of Hoover, Skinner, and Chalmers being satisfied, *and* that there is a unique, ultimate good in maximizing the preferences satisfied among *all* of them. For the majority of mainstream economists schooled since the onset of ordinalism, it is only the former claim that we can make.

7. As I've indicated elsewhere, Jon Mulberg gives a good critical account of the transition from marginalist to ordinalist measures of welfare in modern economics. See Mulberg (1995:Chapter 2).

8. Russell Keat (1997) argues that this way of characterizing "values" and "preferences" is mis-specified insofar as what Sagoff describes as "preferences" can, no less than "values" and "judgements," be rationally defended. When asked to justify one's preferences, one can offer reasons as to why one believes something is desirable, whether its properties actually contribute to this perceived desirability, and whether the good contributes to one's own well-being. Keat, however, acknowledges that policies relying on WTP measures to "value" environmental goods still commit a category mistake by confusing one's ethical judgments with judgments about one's own well-being.

9. That beliefs about what the state of the world ought to be are distinct from beliefs about *how we are to weigh people's beliefs about* what the state of the world ought to be is the primary reason that the argument from justification could not take the following line of reasoning, however appealing it might otherwise sound: "If peoples' willingness to pay for environmental goods were sufficient to establish coherently and sensibly the value of those goods, this would be irrelevant, since whether or not we use WTP measures should itself be a matter of people's willingness to pay to see value established in this way. Hence it wouldn't matter if CVM is 'better' or 'worse' than alternative methods, since its very adoption as a policy tool would be a matter of people's willingness to pay that it be adopted. Yet the question of whether it should be a matter of WTP to decide if we ought to use WTP to establish the value of nature is itself subject to the question of how much people are willing to pay that it be so, and so on in a seemingly endless regress of appeals to WTP. 'Willingness to pay' can thus never itself be justified as an appropriate means of establishing value." This rendering of the argument would thus suggest that allowing for WTP measures at any stage—at any level of a policy question's abstraction—would collapse into a series in which every attempt to justify the use of WTP must itself be justified by WTP. I think the distinction immediately at hand here suggests that such a regress is not necessary, and that one can (and economists typically do) hold that the value of environmental goods should be determined by aggregate WTP, but that this question itself should be decided on its merits. And this is precisely because the beliefs we have about states of the world are independent of beliefs we have about what others ought to believe about states of the world (and, presumably, the grounds on which they might properly hold those beliefs). Whether or not I hold wilderness in high esteem is not the same matter as how I believe we should decide upon the means by which we gage our collective "esteem," and it would be perfectly sensible for me to believe that wilderness should be allocated according to our aggregate WTP but that *this* preference should not itself be subject to WTP.

10. Kahneman's comments are among those included in Chapter 11 of the Cummings, Brookshire, and Schulze (1986) volume, wherein the thoughts of several economists convened for an "Assessment Conference" on the state of CVM in 1984 are collected.

Chapter 12

Strategic Bidding and the Burdens We Wish Upon Others

Let's return to the example of the old-growth forest we have employed at various points throughout Part II. Suppose that the forest is publicly owned, and that the local government wishes to survey the citizenry to assess its "value" for purposes of policy-making. They ask each taxable citizen what he or she is willing to pay per year in increased tax burdens in order to see the forest preserved in its "natural" state for another year. If, the government announces, the aggregate WTP amount offsets the costs of maintaining the old-growth forest—both the opportunity costs of land development foregone and the more immediate costs of pollution control to preserve it's healthy state—the forest will be maintained. Three conditions are thus implied here:

1. There are no guarantees that the good (preservation of the forest for another year) will be provided; its provision is contingent upon a sufficient aggregate willingness to bear its costs on the part of the citizenry.
2. What one "bids" in response to a CVM survey (or any similar means of eliciting WTP assessments) is not necessarily what one would pay if and when the good is ultimately provided. If the governing body finds that the aggregate WTP of the citizenry is more than sufficient to cover the costs of maintaining the forest for a year, they can divide the marginal tax burden in such a way that some individuals end up paying less than they had been prepared to pay. Conversely, some individuals under these circumstances may be required to pay more than they had bid in their own WTP assessment.
3. Individuals who would like to see the good provided will also want to minimize the amount they will have to pay for it to be provided—they will, in other words, desire to bear as little of the burden of its provision as is consistent with its actual provision. We need not attribute purely selfish

motives in this assumption; as we saw in Chapter 10, the rational altruist is interested in getting the best bang for her charitable buck, and would presumably also like to avoid overspending in her contribution to public, environmental goods when that surplus can be put toward other noble goals. Similarly, this assumption need not entail that all individuals will seek to shirk what they believe to be their proper responsibility toward the provision of the good; the propriety of what one believes is a *fair* distribution of burdens—and hence his appropriate share—is precisely what is at issue here.

If we consider these conditions as well as the advice of economists who endorse CVM surveys, we would anticipate that two overriding concerns would weigh in on each citizen's decision when confronted with the question, "What are you willing to pay in increased taxes to preserve the old-growth forest for another year?":

(P1) The probability that one's bid will influence the provision of the good
(P2) The probability that one's bid will influence the amount one will actually be required to pay

The fact that these two leveraging conditions are also assumed to be orthogonal (to vary independently of each other) allows us to further anticipate particular response patterns based on the relative combinations of the two. Specifically, there are conditions under which we would expect particular types of *strategic* responses to the question—responses in which one offers not one's "true" willingness to pay for the good but rather a bid which she believes will influence to her advantage its provision or the amount she will have to contribute toward its provision.

A. If one perceives P1 to be high and P2 to be low, one will tend to *overpledge.* In other words, if one believes that upping one's bid will act to ensure the provision of the good (making it appear more "valuable") while one also believes there is only a slim chance one would actually be held to this bid when it comes time to pay one's taxes, one will tend to state a higher amount than what one's "true" WTP would be under conditions of ignorance about P1 and P2 or their presumed concordance.
B. If one perceives P2 to be high and P1 to be low, one will tend to "free ride," much like the free rider described in Part I. In other words, if one believes that one is likely to be required to pay at least as much as one bids for the provision of the good, and if one believes that one's bid will have little impact on whether the good is actually provided anyway (as in the case when there are many bidders among which to divide the good), one will

tend to state a lower amount than one's "true" WTP under conditions of ignorance about P1 and P2 or their presumed concordance.

Figure 12.1 might help illustrate how these strategic incentives derive from the intersection of P1 and P2:

Randall (1986:117) describes two extreme cases which, while improbable in the real world of public policy, help to illustrate the contrasting conditions as I've illustrated them here. Suppose, on the one hand, one perceives P2 to be zero—that is, that one knows with certainty that her bid will have no influence whatsoever on what she will eventually be asked to pay. In such situations, if one has any positive affinity toward the good, one might be expected to grossly exaggerate one's WTP in an attempt to inflate the good's "value" and thereby ensure its provision; even if one's perception of P1 isn't very high, and one has no overriding reason to believe inflating one's bid will make much of a difference in the provision of the good, there is no risk in such *over-pledging.* Alternately, suppose one perceives P1 to be zero—that is, that one's bid, however large or small, will have no bearing on its provision. In such situations, we might believe one's overriding concern to be to minimize the amount one would have to contribute toward the good and thus, regardless of the probability that one's bid and one's actual burden will be closely linked (P2), pledge an amount under their "true" WTP and *free ride.* In the former case, it would seem that one's only concern is to influence the provision of the good with one's bid; the countervailing concern in influencing the amount one will actually

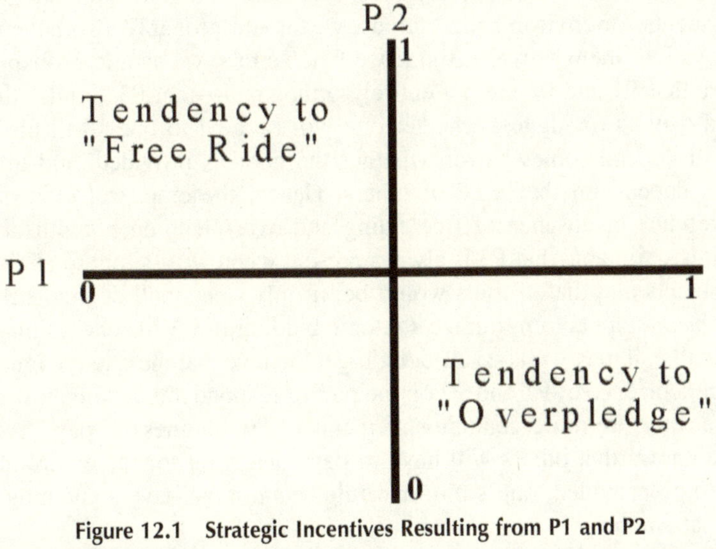

Figure 12.1 Strategic Incentives Resulting from P1 and P2

have to pay is taken out of the picture. In the latter case, it appears that one need only focus on minimizing the amount one would have to contribute toward a promised good; nothing one can do here will ensure or jeopardize the provision of the good.

The operating assumption for CVM surveys, then, is that it is necessary to minimize the possibility of strategic bidding by shielding the respondent from the true values of P1 and P2 (thus keeping their perceived values as close to .5—or pure chance—as possible) or, at least, to convey that their values, whatever their magnitude, are likely to be proportional to each other (rather than inversely proportional, as they are perceived by the over-pledger and the free rider).[1] Is it reasonable to believe, however, that P1 and P2 are truly independent, and thus that there are conditions under which WTP assessments of public goods may be presumed to be non-strategic? If whether I pay what I pledge depends at least somewhat on whether the good is provided (since, in this setting at least, I presumably pay nothing if it isn't), and if whether the good is provided depends somewhat on the amount that *others* pledge, and if, in turn, my pledge depends at least partially on what I *believe* others will pledge, it would appear that P1 cannot be said to be entirely independent of P2. Indeed, we might suggest that the economist must consider yet a third overriding concern that weighs in when one assesses one's WTP for a public, environmental good:

(P3) The probability that others will pledge a sufficient amount to provide the good without a contribution from oneself

If we believe that one's perceptions of P1 and P2 play into one's WTP assessments for environmental goods, we should probably also believe that P3 plays into them as well. And if we believe this, we should probably also believe that P1 and P2 are not entirely orthogonal—that P3 "binds" them in such a way as to suggest whether one will be held to one's full pledge or not will depend somewhat on whether the good is provided, and this will always depend on the WTP of others. Hence, the empirical evidence for the presence or absence of free riding and over-pledging notwithstanding, I think it's arguable that P3 is always present when we attempt to elicit WTP assessments and that all bids would be, if only to a small degree, strategic. While the attempts to minimize strategic bidding in CVM surveys may well be valuable, it may well be self-deluding to believe that the relative ignorance to or concurrence of P1 and P2 on the part of respondents is somehow ensuring that bids are more genuine statements of "willingness to pay." As long as one knows that others will have to pay *something* for the environmental good to be provided, one's bid, it would seem, must have something to do with that amount.

If it's the case, then, that one must consider P3 in one's WTP assessments, and if all bids are thus "strategic" in this minimal sense, what would constitute a coherent bid? If, in other words, P1, P2, and P3 will inevitably be interdependent, how can one settle on a single amount that would represent one's "true" willingness to pay for an environmental good? The economist's account here would seem to hinge the answer to this upon an empirical question: can one make a reasonable assessment of what others are likely to be willing to pay?[2] I think it can be argued that the empirical content of this matter is at best a peripheral question in formulating such a bid, and that for such a bid to be coherent, it must be based on the normative question of what one believes others *ought* to contribute toward the good (or, at least, ought to be willing to contribute). And if such a claim could indeed be defended, it would cast serious doubt on the understanding of environmental "value" as an aggregate of subjective, pre-normative preferences, since individual WTP bids would themselves be the product of moral scrutiny, however implicit.

What are the various beliefs that one might hold concerning how the costs of providing public, environmental goods ought to be distributed? A few candidates seem prominent as possible guiding principles:

1. *According to Desire*—One might believe that the burden should be distributed according to the strength of each individual's affinity for the good, such that those with a stronger desire for the good would pay a larger share of the total cost than those with only a moderate desire, and an even larger share than those with no desire for the good at all.
2. *According to Strategic Parity*—One might believe that each should pay no more or no less than she believes others who also desire the good will pay. The point here would seem to be that no one should be a "sucker" or put other good-desirers in the position of being a "sucker." Thus while this principle doesn't address the relative distribution of actual bids, it does suggest that "sucker" aversion should be evenly distributed. It also suggests that one's comparison need only apply to other good-desirers since, presumably, it would seem fair for one to pay more for the good than someone who has no desire for it.
3. *Equally*—One might believe that the total burden for dividing the good should be divided such that each pays an equal share. Note that some individuals may arrive at the situation with more wealth than others. Hence an ostensibly *equal* contribution might constitute an *unequal* share of the burden.
4. *Equitably*—One might believe that each should pay a flat or even progressive percentage of one's wealth toward the provision of the good. While some individuals would pay more than others, this might correct for apparent inequities suggested by principle (3).

Now my point here is to argue that the "preferences" expressed when one offers a WTP amount concerning public environmental goods are not the primitive, pre-normative preferences we might have for one consumer good over another. One might be tempted to rush the point and suggest something like the following: "Principles (1) and (2) seem to correspond to the types of beliefs that economists suggest individuals might hold going into a situation where the provision of a public good is at stake (though one would have to be a naive economist to adhere to the non-strategic first principle). If these represent beliefs each citizen might hold about the 'proper' distribution of burdens in the provision of an environmental good, we have already established that each must have a normative belief of this sort before she can formulate what she is willing to contribute toward the provision of the good. Just as importantly, we have already established that the economist would readily concede this point. Even if that belief is as simple as 'we ought to pay according to how badly we want it' or 'I ought not get shafted in the deal by paying more than the others,' such a belief would be a normative foundation for whatever WTP value any of the five individuals might eventually offer. If the economist can concede this, she must concede, as you are suggesting, that any WTP value for an environmental good must presuppose at least a minimal amount of normative reflection. We are essentially done here."

We can anticipate how the economist might respond to this line of reasoning. "You may well compel me to concede that one must base one's WTP assessment on some normative principle, even if this principle suggests only that people ought to pay according to their desires or in such a way as to avoid being a 'sucker'. And you might conclude from this that, since normative foundations of this sort need not be present with conventional WTP assessments for consumer goods, the case of public, environmental goods is a special one. There is, however, an important difference in the types of normative principles that you've laid out—at least one of which, you suggest, must be at work in one's formulation of a WTP assessment. I will concede that one might need to hold a principle such as (1) or (2) in order to offer a WTP amount, and that this is more or less implied in the welfare-economic theory behind CVM and the notion of 'willingness to pay' in general when applied to public goods. But note that these principles, unlike (3) or (4), do not presuppose that the good is uniformly desirable or that it ought to be provided. The latter two principles (as well as a host of similar ones we might suggest) answer questions like: 'Assuming the good in question is one that everybody wants or that we might suggest that everyone ought to want, how should we go about distributing the burden of providing it?' The first two principles—the ones you're suggesting I must concede—on the other hand, make no such assumptions about the desirability of the good or about whether it ought to be provided, the inclinations of each citizen notwithstanding. They address the much simpler question: 'What kind of contribution, if any, should one make toward the good?' No *a priori* claims are implied as to whether the good should indeed be provided; whether or not this is the case depends entirely on the aggregate demand for it indicated by the resulting WTP assessments. Yes, I can concede that individuals might need to hold a normative principle of the sort (1) or (2) suggests. This, however, is fundamentally compatible with the belief that public, environmental goods should be allocated like private goods: only to the extent that individuals are willing to pay for them, and not on the basis of a meddling political process which has some individuals telling others what they ought to desire and pay for."

I think the economist's response here is a compelling one. It does indeed seem possible to acknowledge that (1) and (2) can be implied in one's WTP for an environmental good without fundamentally altering the meaning of "willingness to pay." I think the better argument we might construct here would suggest something different; it would suggest that one cannot, regardless of the normative status of principles such as (1) and (2), coherently hold them *and* offer a WTP value. This would mean that not only do WTP estimates for public environmental goods presuppose normative beliefs concerning how the burden of providing them ought to be distributed, they presuppose *particular kinds* of normative beliefs—beliefs more along the lines of (3) or (4) which do indeed consider not only the fairness of the distributed burden in providing the good but also whether the good itself *ought* to be desired.

Think of the matter this way: If we believe that the question of what others are willing to pay in the pursuit of environmental goods is strictly an empirical, strategic one, it might be possible to formulate our own "willingness to pay" for the good. If, however, we concede that whatever beliefs one might hold at this point must try to answer the normative question of what others *ought* to pay, it becomes decidedly more difficult to conceive of how one might settle on any particular amount. Believing this does not require that we make any claims about whether (3) or (4) or similar principles are defensible—this is a separate question of distributive justice. Similarly, it would not be necessary to speak to the normative question of which and when environmental goods ought to be provided (regardless of individual desires)—this would be a question of environmental ethics (one might well believe that the good is not worth providing, and that the "fairest" way to distribute the burden of providing the good would require that everyone pay nothing; this is compatible with the claim suggested here). The point is that individuals must have certain convictions on these matters before they can formulate any coherent WTP for environmental goods. And it is precisely because of this that we would be able to say that the provision of environmental goods is not a question of subjective economic "preference" and our pure "willingness to pay" for them but rather a normative question—of environmental ethics, distributive justice, or whatever labels seem appropriate. Thus whether or not we can solicit WTP amounts from individuals and receive reliable responses is a secondary matter; it presupposes a political process in which WTP "preferences" can be formulated for each individual in the first place.

How, then, might we demonstrate that one cannot hold coherently principles of the (1) or (2) sort and still offer a WTP assessment for an environmental good? It will help us to simplify the old-growth forest example somewhat and stipulate that the community which stands to benefit from its continued existence—and on whose behalf government decisions are made—consists in only five individuals. Suppose further that the governing body assesses that it would cost about one thousand dollars to preserve the forest for another year (based on the costs of protection from fire and other hazards as well as the opportunity costs of foregone development).[3] Suppose also that one of the five—we'll call him Kearney—has at least some desire that the forest be preserved untouched for another year. What are his options if asked "What are you willing to pay to help ensure the provision of the good?"?

One option for Kearney, if he really desires the preservation of the forest, would be to bid the entire $1000, thus offering to carry the full burden of the good's provision himself. It would not be possible, of course, in most genuine policy situations for one to realistically bid the full provision costs of a public, environmental good (unless one were especially wealthy), but since our concern is with what one can believe coherently rather than with what one can do, we can put this concern aside for now. Among the first things that we might notice about such a bid on Kearney's part is that it is incompatible with (and thus could not be prompted by) principle (1). To believe this first principle, Kearney would have to believe that there may be others with desires similar to his who might also believe it is their duty to pay for the good's provision in accordance with their desire for it. But Kearney would also have to acknowledge that his willingness to shoulder the full burden of providing the good would effectively dispatch the obligations stipulated by (1) and let other similarly inclined citizens off the hook; the others who desire the preservation of the forest would not need to bid in accordance with the intensity of their desires and could thus potentially pay the same amount as those with no inclination for the good whatsoever—nothing.

In acknowledging that, with such a bid, Kearney could eradicate the obligation that others pay in accordance with the strength of their desires—and that he thus couldn't coherently hold (1) to be true and be willing to pay the full $1000—we would be implying that principle (1) should be thought of as describing *non-defeasible* obligations. We would be suggesting, in other words, that for (1) to be true as a normative principle, it would have to apply to everyone regardless of the inclinations and actions of some individuals who might be willing to take on a burden larger than that required of them. This would seem to be a reasonable attribute to assign to (1). For if we were to allow for defeasibility in (1), we would need to be able to distinguish required bids from supererogatory ones, and this would require that we are able to point to an amount for each individual that would be required of her in contributing toward the good—even if that amount were zero for those who had no desire for it. We would thus be presupposing precisely what we are trying to determine. To put it another way, we couldn't coherently say that principle (1) would allow Kearney to offer the full $1000 in the preservation of the forest for another year since he's just bidding "beyond the call of duty" if we don't know what amount would constitute his duty. Distinctions between "duty" and "supererogation" demand that we have a reasonable grasp of the line between them; we don't in this case, and thus wouldn't be able to attribute Kearney's bid of $1000 to the latter. It thus appears to stand that bidding the full cost of the good would be incompatible with principle (1).

What about the second principle (2)? Could this be used to justify a bid for the full cost of preserving the forest for another year? This appears to be even less conceivable as a justification than principle (1). Kearney couldn't both pay the full $1000 to ensure the good's provision *and* "pay no more than he believes others will pay" since, if Kearney were true to his bid, no one need pay *anything*. By assessing the value to him of preserving the forest at the full cost of doing so, Kearney would most assuredly be rendering himself a "sucker," and this would be in violation of principle (2).

So far we have established that valuing the forest at its full cost of $1000 is not going to be compatible with either principle (1) or principle (2). What about the other extreme? What if, in other words, Kearney were to assess the value of the good to him at zero, in spite of the fact that he has the same amount of desire for the good as he did when we considered the full $1000 bid? Would principles (1) and (2) fare any better under these circumstances? Not likely. It seems clear, on the one hand, that principle (1) would be no more compatible with a zero bid than it would be with a full $1000 bid. If, as we've stipulated, Kearney does have some desire that the good be provided, principle (1) would require that he want others with similar desires to pay the same amount he is willing to pay, and that others with a lesser desire pay less for the good. But this wouldn't be possible. Even if Kearney could will that others with the same desires as he has for the good pay the same amount—nothing—he couldn't will that those with a less intense desire for the good pay less, since they can't pay less than nothing (they could, conceivably, pay a "negative tax" analogous to an earned income tax credit and receive a check from the government, in which case their "bid" would be a willingness-to-accept amount that would sufficiently compensate them for their tolerance in bearing with the provision of the good). It appears that if Kearney were to sincerely hold (1) to be true, and if he has at least some desire for the good, a WTP bid of zero would not be consistent with these conditions.

It might seem, on the other hand, that principle (2) is compatible with the zero bid option. If Kearney believes it is his duty to avoid being a "sucker" by paying more than he thinks others might pay, assessing his WTP at nothing would be about the safest bid he could offer. It might be tempting to counter this, however, by appealing to the effects of Kearney's zero bid on other bids. The second principle stipulates that one should not render one's self nor *anyone else* a sucker—that is put oneself or other citizens in a position where one would have to shoulder a greater share of the cost than they believe others are willing to shoulder (regardless of the magnitude of the bids). We might think that if Kearney were actually to bid nothing when prompted for the value of the forest to him, he would be putting anyone else who might desire the good and who thus might be prepared to pay something toward its provision in a position of bearing a heavier burden than she would have to bear otherwise. As morally problematic as this might sound, it would not be in violation of principle (2), which concerns not the actual burdens others bear but the burdens one *believes others are willing* to bear. *Strategic parity*, in other words, requires not that Kearney pay any particular portion of the total cost of providing the good, nor any particular amount in relation to his fellow citizens. It requires only that he not pay more than he thinks others will pay *and* that he not put others in a position of having to pay more than they think Kearney and others will pay. In bidding nothing, Kearney has no reason to believe his four fellow citizens are aware of his bid; while they might, if they have no more desire for the good yet are willing to pay *something* for it, be "suckers" in the eyes of an impartial third party, they are not knowingly putting themselves in the position of being a sucker by virtue of Kearney's zero bid. Hence this possibility alone doesn't render Kearney's zero bid incompatible with (2).

A different kind of argument, then, would be required to show that a zero-bid from Kearney would be incompatible with principle (2). One additional route may be available. In order for us to understand principles (1) to (4)—as well as the numerous other principles we might consider which purport to stipulate a "fair" distribution of the burdens in providing environmental goods—as *moral principles*, each would need to be *universalizable* in a rough Kantian sense. They would, in other words, need to meet the condition that, in believing the principle to be true and to be something to which one should adhere, one would also need to believe that everyone can and should adhere to it (much like the contrasting beliefs of the free rider and the honest holdout we saw in Chapter 6). Principle (2), one might suggest, cannot meet this universalizability criterion in conjunction with a zero bid. One cannot, in other words, desire the good, bid nothing toward its provision, hold (2) to be a compelling moral principle, *and* coherently believe that everyone else who similarly desires the good can bid nothing and adhere to (2). They too would reach the same false conclusion that a zero bid is compatible with the provision of the good and a "fair" distribution of its burdens (in terms of Strategic Parity); they would thus also bid nothing for the good, and all would see their intentions in seeing the good provided go defeated. Note that, despite superficial similarities, this concern is different from a Tragedy of the Commons scenario and the n-player prisoners' dilemma, as discussed in Chapters 1 and 2. This is not a situation of counterfinality, where the self-interest of each is undermined in the aggregate. This is a situation where rational, uniform adherence to a normative principle systematically undermines the ends of the principle. In theory, all five citizens in our example could desire that the good be provided and could bid nothing in seemingly perfect accordance with the principle of Strategic Parity. This would, however, ensure systematically that the good isn't provided. Hence principle (2) would seem to be incompatible with a zero bid from Kearney or from any citizen who desired that the good be provided.

We have thus far established that Kearney, in offering a WTP assessment to preserve the old-growth forest for another year, cannot coherently bid either the full $1000 or bid nothing and accept as morally compelling principles (1) or (2)—the two distributive principles which, the economist is quick to point out, make no presumptions about the desirability of the good or about the normative propriety of its being provided. What about everything in between these two extremes? Is it possible, for example, for Kearney to bid some amount $x, such that $0<$x<$1000, in accordance with principle (1)? I think it can be seen that we must answer "no" here as well, and that no matter what amount we substitute for x, it would be incompatible with (1). This is because, as we'll see, no matter what Kearney might think he would bid as a WTP amount, how the rest of the burden would be divided among the remaining four citizens in our example would necessarily be arbitrary.

Suppose, for example, Kearney believes he desires the good substantially more than his fellow citizens and sets his bid at $600. We might think his job here is done and that he need not care how the remaining $400 burden be distributed among the other four. This would be premature, however, since it would be inconsistent with what the first distribution principle (1) requires. The principle suggests that each person's WTP reflect her relative affinity for the good, such that those with stronger desires that it be provided contribute more than those with less pronounced desires. This suggests that any one individual's bid must be established in regard to *everyone else's* bid. In order for one's bid to meet the conditions of (1) in other words, it must be made in consideration of all other possible bids, and it must reflect an amount that represents one's desires relative to the possible desires of all other citizens.

Hence it would not be consistent with (1) for Kearney not to care how the remaining $400 would be distributed if he were to bid $600; his bid must reflect not only his own desires, but what he perceives to be the desires of everyone else.

Now a $600 bid from Kearney will allow for a number of ways in which the remaining $400 of the burden can be divided. For the sake of argument, let's rule out all of the possibilities in which the good would not ultimately be provided and those in which its provision is overdetermined—that is, those in which the total WTP bids by the remaining four citizens sum to less or more than $400. We'll consider, then, only the situation in which the array of remaining bids do indeed sum to exactly $400. Even with this restriction, Kearney must know that there will be quite a few ways in which the $400 burden can be distributed among the other four citizens in our hypothetical world, including four different distributions of $400- $0-$0-$0, as well as a 4 x $100 distribution, and a host of others. In order to settle on a $600 WTP bid, Kearney would have to examine this finite but large array of counterfactual conditions—a daunting task, but one that we cannot say is utterly impossible. The value would be astronomical in the case of real-world calculation of one individual in a large society, and considering each would be, for all practical purposes, impossible.[4] Demonstrating such a bid's practical difficulty is not, however, what we are aiming at here.

Of all these possible distributions, how would Kearney decide which is best in accord with principle (1)? Presumably, he would need to make some appeal to what he believes are the relative desires of the other four citizens. Even here, the possibilities proliferate. He could, for example, assess how strongly each desires the good in proportion to his own stipulated WTP of $600; for example if he believes one of the four to desire it one-third as much as he does, he can assume that this individual's WTP should be $200, while the next most desirous who values it at half that should bid $100, and so on. Similarly, he might simply rank the remaining four in terms of perceived desire and believe that they should contribute progressively smaller amounts toward the provision of the good, with the interval between any two not substantially different from the interval between any other two. The problem with either of these schemes, and with, for that matter, any scheme of which Kearney might conceive in allocating the burden among his fellow citizens, is that their *relative desire* is relative to Kearney as well. Kearney has already restricted the range of possible distributions by arbitrarily setting his own WTP at $600. While there remain many possible ways to distribute the remaining burden, it is a set decidedly different from the possible set that would obtain had Kearney bid $599, or $250, or $996. In setting his bid at $600, Kearney has no way of knowing with certainty that the normatively "best" distribution (i.e., the one that most closely conforms to the relative desires of the five citizens) lies among those available with four desirers and $400. It could just as well be found only among the possible distributions that result from a different bid. *And this would be true no matter what Kearney's bid is*—for any amount x where $0<$x<$1000. Hence there is no amount that Kearney could bid and of which he could say "this bid accords with the relative strength of my desire that the good be provided, and leaves an amount which can be distributed by the same criterion for everyone else." There is, in short, no amount Kearney could coherently bid and conform to distribution principle (1).

Note that the problems Kearney faces in pinning down a WTP bid that would conform to (1) do not constitute a purely cognitive, strategic question of mutual uncertainty about what others are *likely* to bid. The problem holds even if each assumes the other four are acting on principle (1) rather than strategically (i.e., bidding more or less than he really thinks the good is worth to him, if this is indeed possible). It is a moral problem of believing that everyone should pay for the good's provision only in proportion to her relative desire, but being simultaneously unable to establish how WTP bids should act as a marker for that desire. More information—even a lot more information—is not going to be of use to Kearney here. Even if he could have some firm grasp of how the strength of his desire that the good be provided rates in proportion to the strength of others' similar desires, any WTP amount he might attach to his desire immediately forecloses on innumerable other ways to attach proper WTP amounts to desire. It is true that Kearney cannot know with certainty that he selected the "proper" amount that would allow others to bid according to the strength of their desires. The important point is that what he loses here is not a strategy for making the most advantageous bid, but rather a normative justification—assuming principle (1)—for *any* bid.

Thus a partial bid of something greater than zero but less than the full amount that is compatible with (1) appears to be precluded; hence no bid can be compatible with (1). What about the second distribution principle (2)? Is there some partial bid x, where $0<$x<$1000, that is compatible with Strategic Parity as we've been describing it? If there were, it seems that much of Kearney's WTP decision would be determined by how many among his four fellow citizens also desired the good. If he assumes that each of the other four has some desire for the good and that they are each bound by (2), Kearney's bid would have to be $200—the only bid any of the five can offer and be assured that no other good-desirer and (2)-adherer is bidding less. Any bid larger than $200 would run the risk of making Kearney a "sucker"; any bid smaller than $200 puts the provision of the good in jeopardy. For the same reasons, if Kearney were to assume that only three of the remaining four citizens desire that the good be provided, his maximum bid would be $250, since a larger bid would decrease the total burden the other three good-desirers would have to shoulder, allowing at least one of them to bid less than $250 (and again rendering Kearney a "sucker"). Likewise, if Kearney assumes only two of his fellow citizens desire that the forest be preserved another year, he would have to bid $334 to be within the bounds stipulated by the second principle (we'll round up the actual $333.33, since we've been dealing only in whole dollar units thus far). If he believed only one other citizen desired the good, $500 would have to be his bid, and if he believed he were alone in desiring that the forest be preserved another year, the same reasoning would have him bidding $1000, at no risk of being a "sucker."

We've already ruled out the possibility of a $1000 bid consistent with principle (2) on other grounds, so we can safely assume here that if Kearney is going to appeal to this distribution principle in formulating his WTP bid, that bid is going to range from $200 to $500. If we assume any particular level of desirability among his four fellow citizens, we can further pinpoint that amount. If, for example, Kearney has good reason to believe all four of his counterparts desire the good, $200 would be his best bid consistent with (2), as it would be for the others. Whatever assumption he makes about the good's desirability and thus the relative compliance with (2) among the others, he would be able to formulate a WTP amount consistent with (2).

It thus appears that we've achieved some rational, coherent WTP value with principle (2) alone, and that my prediction that this wasn't going to be possible has been undermined. Is this the case, though? Note that whatever value Kearney can settle on in this example—$200, $250, $334, or $500—will depend on how many of the other four citizens he believes have some desire that the good be provided. Whether, in other words, he can settle on $200 or whether he believes he must up his bid to $250 depends on whether he thinks the burden will be properly divided among 5 or 4 persons (and whether he should raise his bid still higher depends on whether he believes the burden will be divided among still fewer good-desirers). This might be, on the one hand, a decision of Strategic Parity based on available evidence and his assessment of how many fellow citizens are likely to desire the good and thus, under the weight of (2), reach the conclusion that they too should bid an amount that leaves no one a "sucker" and yet doesn't jeopardize the provision of the good. It might, on the other hand, be a normative decision about whether the good ought to be "desirable"—an affirmative answer to which would suggest that all should bid the $200 that (2) would suggest. How can we know which kind if decision it would actually be?

Consider, more particularly, just one of these decisions—the strategic decision as to whether one other person is likely to desire the good and be willing to contribute (i.e., whether Kearney should bid $500) or whether it's likely that two other individuals might desire the good (making his required bid $334). How is Kearney to arbitrate between the two options?

He could try to assess the "tastes" of the others based on available behavioral evidence. This isn't likely to be conclusive however, since people with similar "tastes" can exhibit different behaviors depending upon how they understand the things they believe to be satisfying them. "Revealed preferences" of this sort, moreover, are "revealed" to the economist by virtue of market behavior; we believe we are assessing one's "tastes" by witnessing what one purchases more of and what one purchases less of, or not at all. This isn't at Kearney's disposal here, since there aren't yet markets or prices for the good in question. This is precisely what we are asking Kearney and the other four to establish, and we can't appeal to the presumed cornucopia of behavioral information markets provide before they even exist. On the contrary, in order to decide whether the good is likely to be desirable to any additional person, it seems Kearney must examine what it is about the good that makes it desirable to him in the first place. These qualities are not going to be, as they might be in the case of consumer goods, matters of pure "taste" but rather beliefs about the benefits the provision of the good would bring to everyone, or about the intrinsic benefits of the goods existence.[5] Hence if Kearney can hold these beliefs himself, he must believe that another can hold them, and another. It is because of this that we might expect him to reach the conclusion that if (2) is to be their guide, the strategic decision as to whether he will likely be contributing with one or with two other individuals will be resolved in favor of the latter.

But if Kearney can believe that this would justify assuming the existence of a third good-desirer, he must believe that it justifies a fourth, and a fifth. If the argument applies to any marginal good-desirer, it must apply to all. And if this is true, it appears that what initially presented itself as a strategic decision about how many good-desirers there are likely to be among the other citizens collapses into the normative decision about whether or not each citizen ought to consider the good desirable. And if *this* is true, it would appear that Kearney has moved seamlessly from justifying his bid with the second distribution principle (2) to justifying it with principle (3)—one that, to the economist's chagrin, requires Kearney to have normative beliefs about the desirability of the good to all individuals before he can formulate a WTP value. Thus, unlike the preceding arguments which attempted to show the incompatibility of principles (1) and (2) with various bids, the problem with any partial amount x, where \$x<\$0<\$1000 is that any attempt to justify such a bid with (2) is ultimately going to dissolve into a justification by (3); principle (2) alone cannot justify any particular bid.

I've spent the last several pages arguing that various possible bids are going to be incompatible with principles (1) and (2) as laid out above. Table 12.1 provides a shorthand summary of how each alternative was ruled out.

Table 12.1 Summary of Incompatibility of Possible Bids with Distributive Principles (1) and (2)

Bid	*Principle (1)*	*Principle (2)*
$0	Must will that others with same desire for good bid the same; hence inconsistent with provision of the good	a) Would render other good-desirers "suckers" b) Not both universalizable and consistent with provision of the good
$0<$x<$1000	Cannot non-arbitrarily select a bid and choose "best" distribution of remaining bids among other good-desirers	Any strategic rationale for assuming x good-desirers collapses into normative belief that all others should desire provision of good—i.e., principle (3)
$1000	Displaces obligation of other good-desires to contribute in accordance with their desires	Renders himself a "sucker" if there exist other good-desirers prepared to pay

The purpose of this exercise was, of course, not to defend any particular principle such as (3) or (4), but rather to argue that if, as the economist might suggest, there is a difference between distribution principles which presume the desirability of a good and those that don't, the latter aren't going to be workable principles that individuals might use in formulating WTP amounts for particular bids. Truth be told, I could probably also argue that bidding the full amount and bidding nothing are no more compatible with (3) or (4) than they were with (1) or (2). The difference is that bidding *some partial amount* of the cost of providing an environmental good is indeed compatible with principles of this sort. Note also that I haven't said anything about the options for those who have *no desire* for the good at all. These individuals, conceivably, could hold either (1) or (2); this would be consistent both with their desires and with the actual non-provision of the good. These principles are not, however, something they could will everyone to hold; as we've already seen, those who desire the good can't hold them, and one can't coherently will that someone hold a moral belief that it would be impossible for her to hold consistently with her other beliefs and desires. Hence it would be possible for one to hold (1) or (2) only in the odd chance that *nobody had any desire* for the good in question. It might thus appear that we have a case in which the provision of the good was a non-political, consensus issue, and might imply that the provision of the good is resting ultimately on subjective inclinations. This would be inaccurate, given what I've argued above, since we would have to understand the lack of desire for a public, environmental good and the attendant willingness to pay nothing for it not as an expression of consumer taste but as a judgment about whether *anyone* ought to desire that the good be provided. There would be nothing contradictory about universal adherence to (1) or (2) provided a reasonable belief that the good in question (perhaps the Ebola virus) is something no one would desire.

The reasonable person will, no doubt, respond to the arguments in this chapter with something like the following: "Good grief; It was hard enough to follow the tedious lines of reasoning as you presented them. Am I to understand that individuals, when prompted with the simple question 'What are you willing to pay for environmental good x?' go through this regimen of self-examination, assessing their normative beliefs and their stipulated WTP values for consistency, and reaching the conclusion that, whatever value they might decide, it is one based on moral principles that they may not even be able to articulate? The 'Methods' sections of the CVM research literature don't seem to support this account. It seems to more typically go like this: we ask the question; subject answers, or subject doesn't. End of story."

Commonsense concerns about just how deep the waters of moral psychology run are probably never entirely misplaced, but they miss the point in this case. The claim here is not that individuals go through this painstaking process of eliminating their options and examining their moral beliefs each time they are presented with the opportunity to assess the value to them of public, environmental goods. The claim, rather, is that they cannot *in principle* find that they maintain the types of beliefs economists would attribute to them even if they tried, and if they thought that they *could* maintain such beliefs and offer a WTP bid, that bid might indeed need to be treated as nonsense.[6] In many ways, the economist's "story" of how we go about examining our "tastes" and offering up WTP bids for environmental goods sounds no more psychologically plausible than anything I've described here merely by way of illustration. To describe such values as primitive, pre-moral "preferences" ask us to believe that individuals dip into some inarticulable registry that assigns numbers to different levels of environmental good provision. Even ordinalist economists who might otherwise eschew the presumption of "utility" would have us believe that selecting one public good over another will tell us something about what the selected good "does" for the individual in terms of preference satisfaction. Clearly, the issue here can't be one of who has the better psychological story to tell. The issue is one of what we can reasonably believe about the moral content of the values that get expressed.

This leaves us with the claim that, if individuals are to have any meaningful sense of what a particular environmental amenity is "worth" to them, that value must derive from normative beliefs about whether the good ought or ought not to be provided and the means of fairly distributing the burden of its provision if it indeed ought to be. There are no answers to these questions etched in stone. There are, however, better and worse forms of political discourse and their corresponding institutions which we can employ in establishing their answers, and the "marketplace"—while *possibly* useful later in the process of enacting particular environmental policies—doesn't appear to be one that can be useful in establishing the value of the environment in the first place.

NOTES

1. Mitchell and Carson (1989) devote a full two chapters (Six and Seven) to the question of strategic bidding in WTP assessments, including a useful review of the experimental literature in which contingencies such as P1 and P2 above are controlled in a laboratory setting. In general, the authors concur with those researchers who have come to the conclusion that, empirically, there is insufficient evidence to believe that strategic bidding systematically undermines the reliability and validity of WTP assessments, and that it can be mitigated with relatively little extra effort in CVM methodology.

2. This is largely the area economist refer to as the problem of *demand revelation.* Edward Clarke is among those most prominently associated with demand revelation as it pertains to the provision of public goods. See Clarke (1980) for an overview of his approach as well as Scherr and Babb (1975) and Clarke (1975) for an experimental application that tests Clarke's pricing system against those of some of his competitors.

3. In the notation introduced in Part I, i_k would thus be the individual whose bid puts the aggregate WTP over the $1000 mark. P3 above, in this scenario, would thus be the assessment by each of the five of the probability that i_k is among the other four.

4. The general problem is one among those in mathematics know as *combinatorial* problems, and amounts to the question of how many ways we can distribute *p* objects among *q* containers (assuming the "objects" here—dollars—are irreducible units). The formula,

$$\frac{(q + p - 1)!}{(q - 1)!\, p!}$$

hints that even relatively small values for *p* and *q* will result in a substantial number of potential distributions. Even in Kearney's simple world, if only ten dollars were to be distributed among the remaining four citizens, this would leave him with 286 possible distributions he would need to consider. Increasing this to twenty dollars—still well below the $400 in our example—increases the number of possible distributions to 1,771. This would seem to confirm that real-world situations where larger societies and more costly goods are at stake would entail numbers that are unfathomable for most individuals. For a demonstration of combinatorial and similar mathematical principles, see Feller (1968:Chapter 11, especially 38–40). I thank Thomas Leuthner for helping to clarify the essentials for me.

5. Note, as I have before in similar contexts, that the quality or normative defensibility of these beliefs is not what is at issue here. They could well be misguided, indefensible, and downright strange and the type of decision Kearney would be making would remain the same.

6. One might object here: "You have already that economists need not be wedded to assumptions of egoism in understanding individual preference satisfaction and that there is no inherent conflict between altruistic motivations and economic preferences. Now you seem to suggest that these things are at odds or, at least, that there's really not much by way of 'preference satisfaction' going on in these matters. Which are we to believe?" Despite appearances, there is really no contradiction here. The previous claim was that altruistic, charitable inclinations could indeed inform otherwise "rational" market preferences without "tainting" either; this seems perfectly compatible with the current argument. The earlier concerns were with what individuals understand to be voluntary, supererogatory "contributions" (via market actions) to a cause. The argument here is that, in the case of public, environmental goods, assessing their *value* as such requires us to evaluate what others *ought* to contribute to this effort. The fact, in other words, that altruism is compatible with market preferences does not undermine the fact that what we believe others are *obliged* to do must enter into the value we place on public goods in the collective schemes which must be organized to provide them, and that this must be *antecedent to* any "market" value of such goods.

Concluding Remarks to Part II

We need to be cautious about what we might take from the preceding discussions. Many of the critiques of CVM by mainstream economists as well as the critiques of WTP measures in general by environmentalists that were considered here aim to undermine, to one degree or another, the validity of such methods. These claims, for widely disparate reasons, suggest that we are not really measuring anything meaningful when we ask people what a particular environmental amenity is worth to them—that the "values" which they might report have little to do with the "value" of the environment, be it a "price" or some other designation. My intentions here were not quite as ambitious as this. I don't think anything I've said here should cause us to either abandon or embrace the plethora of research that has emerged from CVM studies, let alone the century of welfare-economic thought which exalts "willingness to pay" to the status it enjoys today among economists. Rather, my primary concern has been with the normative content of those methods which purport to return a pure, pre-political assessment of the "value" of a public good—an assessment that is putatively similar in kind to the value assessments we might have for consumer goods readily allocated on a market. I believe we do have adequate grounds for doubting that the fact one might state a "price" for environmental amenities when asked entails that its value may be assessed in the same way as other goods. We should be careful, however, in how we justify this doubt. It is not simply because clean air is intrinsically more "valuable" than canned soup, or that wilderness preservation seems more noble (and thus "invaluable") than a pair of brand-name sneakers, though these things are almost certainly true. It is, rather, because our assessment of what clean air or wilderness mean to us—if we wish or are asked to express this value in terms of what we are willing to pay for these things—presupposes that we examine particular normative concerns about what we can demand of other individuals in their provision, in correspondence to or in spite of the "value" *they* claim to place on the environment. This seems clearly to distinguish environmental quality from things which we typically understand to be marketable.

Does this mean that free-market environmentalists, libertarians, and other advocates of privatizing of the natural environment are simply misguided? If we are to understand these positions as suggesting that economics precedes politics in the provision of public goods—that sound environmental policy is built first around the assessment of brute, pre-normative preferences for environmental amenities—then I do think we have grounds for asserting they are wrong on *at least this fundamental point*. This would not entail that markets are entirely useless in the process of providing environmental goods, or that the relative value of environmental goods in deciding how we spend our social wealth can never be determined by asking individuals what particular amenities are worth to them. It does, however, suggest that these things are part of a larger political process, and that if we believe we're *initiating* the process by soliciting assessments of what individuals are "willing to pay" to make environmental goods available—and especially if we believe this would imply that the natural environment can be readily privatized and sold off to the highest bidder—we are deluding ourselves. Free-market environmentalists and market advocates of all stripes tell us repeatedly that markets for environmental amenities, as in other areas, "get prices right." This claim presupposes that there exist primitive, pre-reflective prices to be had for goods, and that tapping them is strictly a methodological question of economics. Whether or not this presupposition may be a useful one for consumer goods, I have argued in Part II that, when it comes to public, environmental goods, it will lead us to ask the wrong questions.

For many of the same reasons I would counsel restraint in dismissing entirely the economic methods I've examined in Part II, nothing I've argued here would recommend that any particular environmental good should or should not be provided. What I *have* argued, rather, would suggest that the fact that no apparent "market" exists for a particular environmental good is insufficient reason for not considering whether the good ought to be provided—or, to put it in terms less beholden to economic theory, whether particular efforts in making the natural world more viable ought to be pursued. I think the practical upshot of what we have seen implies that basing the "value" of environmental goods at least in part on what individuals are prepared to sacrifice—both personally and socially—to see them provided is prudent, but that this bears little resemblance to going shopping. Unfortunately, given the hold of free-market ideology on public policy and social discourse in the U.S., a make-or-break nation for global environmental policy whether we like it or not, we can have little optimism for understanding the complexities of public good valuation as a unique civic endeavor.

The fact that most of what I've said in Part II, if not the entire book, would likely strike most of the adult population in the English-speaking world as opaque, tiresome, and utterly foreign is in fact the problem the text addresses: we are not prepared by our educational and civic institutions to think of the world of "goods" we might desire and be willing to bear some cost to provide as anything but consumer items in a vast world supermarket. For those of us raised in the U.S., quite literally schooled in free-market economics (if your secondary-school civics-class education consisted in anything but capitalist bromides and tales of the historical "triumph" of the market, consider yourself fortunate), it's all a matter private, marketable goods with a few peculiar things like civil defense, mass literacy and clean air that seem to gum up the otherwise smooth-running system. That libertarian think tanks feel a need to spend millions of dollars annually to keep things this way (by, e.g., touting the virtues of "free-market environmentalism" if not directly funding the evisceration of environmental regulatory policy) suggests how quickly we might unlearn this ideological trope if we allowed ourselves to do so.

It's not, in short, that we cannot pin a "value" on nature—or even that we cannot assign a dollar amount or some other measure of the burden we are personally willing to bear to maintain the planet's viability. I've maintained throughout Part II that a moon-eyed retreat to "don't make me put a price tag on a waterfall" is not the best response to the free-market economist that we can muster; there is no shame in being able to assess what specific aspects of environmentalism are worth to us, what we are willing to forgo as a matter of lifestyle choice or social choice for the sake of the environment, and how much we are prepared to demand that others contribute toward the environmental state of the planet that we aspire to. The point, rather, is that we cannot assess this value in the context of an economic system that limits us to selecting environmental amenities like we select melons in the produce section, and on the (often accurate) assumption that we've put about as much forethought into it.

Similarly, nothing I've suggested here should be understood as an endorsement of either unfettered markets or the state powers that vie to "meddle" in them. While I have conceded at various points that markets might be an effective tool in implementing some environmental policies, and that there is an irreducible authority that may be given over to government in enforcing legitimate cooperation, neither camp in these debates is likely to provide the answers to the environmental questions that we really should be raising, and which I have tried to articulate in Part II and throughout this book. *This* claim, moreover, should not be understood as advocating "capitalism with a human face"—that is, some watered-down version of current arrangements coupled with state intervention to save it from its own vices and steer us clear from the extremes of rapacious private greed and oppressive state-centralized communism. Rather, it suggest that we need to consider other means of organizing political objectives and deciding how we will spend the surplus wealth that we produce in "developed" societies. In simpler political times, we would be able to just call this the "socialist project" and get on with the business of realizing it. Things are complicated now by historical hindsight and our knowledge that a fully planned allocation of all the goods that we can produce will have a great number of problems—some of them economic, some of them ethical. I suspect that we will eventually reach the conclusion that both the democratic state and markets will play a necessary role in whatever political/economic structures emerge, and that we will need to decide via the former when and where the latter are useful and appropriate. The justification for whatever balance is struck would, however, lie not only in its productive efficiency but also in the degree to which it conforms to sound normative principles of fairness. Some call this ideal "market socialism" and have renewed interest in the concept; others shy away from the label, given qualms with one or the other of the elemental terms or the conviction that it is an oxymoron. Whatever we choose to call it, and whether or not the arguments presented in this text suggest it, I believe our ecological, economic, and political future requires that we give it serious consideration.

Afterword

The Poverty of Ecology

If the proverbial alien from another planet were briefed on our current ecological malaise, the creature might conclude from the summary that our condition was the result of a tremendous, valiant effort to share widely the planet's wealth. Surely, the being would infer, the problems you describe come from the fact that you've managed to feed one another, provide each person with a home, and ensure a viable existence for all. You meant well; you simply wanted to get everyone up to speed, providing them with the basic requirements of living. The machinery you put in place just got the better of you. Now that everyone is living comfortably, that machinery is taking its toll on the planet. You just need to lighten up a bit—share the wealth, still, but produce a bit less of it to share.

We must at this point mask our embarrassment and reply that, no, this is in fact not the situation at all. Our plight is not the result of having too effectively eradicated basic want; we still have large swaths of the planet's population living in squalor. The ecological malaise is not a necessary by-product of having met everyone's basic needs but rather a consequence of meeting some people's needs to obscene excess while leaving the rest behind. Instead of swamping the lifeboat in a futile attempt to save the lives of all aboard, it's more a case of us submerging the rabble in the stern in order to keep first-class passengers elevated and relatively dry at the bow for as long as possible. We are, in short, exploiting global resources to maintain a relative handful of extravagant lifestyles. A disproportionate few in the developed world have fed themselves and shopped themselves nearly to death, and they're taking the rest of the biosphere down with them. As Pope Francis has pointed out with characteristic frankness, we can't claim the noble achievement of having eradicated want as our excuse for wearing thin the ecological blanket in which we find ourselves swaddled: "The exploitation of the planet has already exceeded acceptable limits and we still have not solved the problem of poverty" (Francis I, *Laudato Si*: #27).

We would no doubt marshal economic excuses as our defense for the situation. We're trying, we tell the bewildered interstellar traveler, to spread the wealth around—to get everyone on board with the production of sufficient food, the procuring of shelter from the elements, and the prospects of a life beyond a brief, harrowing infancy. We're not all that greedy or callous, we explain. We'd like everyone to have the same comforts some of us enjoy, but haven't yet figured out how to spread those comforts around, and the effort to do so is eating away at the foundation. We realize that this way of life takes a toll on the planet; we're still perfecting ways to lessen the toll while others work to achieve our way of life.

What response might we offer that alien creature when it assesses the landscape of our problems and, inevitably, suggests that perhaps we're going about it backward? Rather than the Promethean task of welcoming the planet's dispossessed into the lifestyle of Western industrialism's comfortable classes, perhaps, for the sake of the planet, they should be welcoming us into a lifestyle a little closer to theirs.

Beginning years before the project you are reading was conceived, I would have found that advice from our alien friend a bit suspicious. I identified myself as a democratic socialist (still do, incidentally, as the reader may have gathered) and only a "reluctant" environmentalist, as attuned to matters of distributive justice as to the occasionally "bourgeois" preoccupation with preserving camping spaces and breathtaking vistas for those with the leisure time for outdoor activities. The middle-class environmentalist's notion that the solution to our ecological ills was to leave the impoverished in their condition and simply emulate it, renouncing her trust funds, her Ivy-league education, and her familial connections to rent a loft space in places like Chicago's trendy Bucktown neighborhood and subsist on rice and chickpeas seemed positively silly and politically inept, if not condescending. Pretending to be poor, I long objected, whatever it might do for the pretender, did little or nothing to address the situation of those who don't have the wherewithal to pretend to be anything but poor, and arguably helps to perpetuate an economy where most have not enough and some have way too much. Self-aware, make-believe poverty, in short, exacerbates the real thing, or at best leaves it entirely untouched (one telling example: E. F. Schumacher's *Small is Beautiful* counted, for some Reagan-era Republicans, as sacred scripture no less than it did for many environmental activists and pioneers of the "minimalist" and "back-to-the-land" movements). If saving the planet meant informing the impoverished that they will never have the opportunities we've had, thoroughly squandered, and are now ready to relinquish for some dreamy idyll of the "simple life," I wasn't entirely sure this was a planet worth saving.

Only later and for reasons I can't fully explain did I find the lost Catholic faith of my youth, and soon afterward entered discernment for religious life as a Franciscan brother and priest. It would be convenient to suggest that the somewhat radical spirit of St. Francis of Assisi had a hand in these developments, but I'm reticent to speak to the ratio of pure chance to divine guidance at work here. More importantly, the decision only heightened the ambivalence with which I'd viewed the environmental movement: by entering the Franciscan order, I would be preparing to take a lifelong vow to a voluntary "poverty" quite similar in appearances to the sort which I'd long dismissed as a mannered, middle-class affectation.

Giovanni Francesco di Bernadone—known to us now as St. Francis of Assisi—was among many other things just plain crazy. He carried out his pieties and devotions well beyond the point of physical self-abuse (admitting as much toward the end of his short life), and asked similar fortitude from those who chose to live as his brothers and sisters. If we are to believe only half of the stories about him that circulated among Franciscans and beyond in the centuries following his death in 1226, we would be forced to conclude that he was a gentle and devoutly religious man who, had he lived in our more clinical era, may well have been forcibly institutionalized to prevent serious harm to himself.

Then again, this is pretty much the point, isn't it? We don't make saints of people with merely a mild affection for God or a tepid approach to life, and people worthy of our emulation—especially devotional emulation—will almost certainly have habits and traits we would find repulsive in polite society. Francis, once we put aside the plaster image atop the birdbath and the Franco Zefirelli, *Brother Sun, Sister Moon* rehabilitation of the lovable proto-hippie, lived on the edge of what we like to think of as sanity, and this inevitably put him on the edge of society, a society that was already writhing to shake off the manacles of medieval institutions that Francis simultaneously embraced and threatened. He was a perfectly contradictory figure for perfectly contradictory times.

Pope John Paul II deemed Francis the patron saint "of those who promote ecology" just shy of a decade into the modern environmental movement (if we mark its beginnings, as many do, with the first Earth Day in 1970), though it would be an anachronism to suggest that Francis was himself an "environmentalist." The term and the movement, as we've seen throughout this book, can only make sense in the context of an industrial or post-industrial society that has asserted its "mastery" over the natural world, wresting from it a reliable means of subsistence via methods which it has ultimately come to question. Still, Francis is credited by his earliest biographers with composing *The Canticle of the Creatures* in his native Umbrian Italian dialect, and the legends surrounding him that would circulate long after those early biographies were suppressed by a Franciscan order bent on managing his legacy would have him commanding the attention and esteem of forest-dwelling creatures, cheerfully silencing birds whose chatter overpowered his preaching, and brokering a peace deal between a besieged town (Gubbio) and the wolf who terrorized it.

Just as important, it was Francis who gave us the model of the voluntarily impoverished—the person of means who renounces it all for the higher good of the Kingdom of Heaven. "Voluntary simplicity," if we may apply our modern term to medieval life, had been in existence for centuries, and religious motivation for such a lifestyle was nothing new. Even among his contemporaries, Francis' message was somewhat blurred by the fact he was acting on impulses already put into practice by a host of other lay groups disgusted by aristocratic and church affluence who chose to live lives of austerity in common, ostensibly at the behest of the Christ of scripture. The difference for Francis was that approximation would not suffice. To live the Gospels was not simply to do as the Nazarene had said and adhere to his teachings but to follow in his footsteps as accurately as is humanly possible. No home to call one's own, no provisions beyond what will sustain one for the time being, and, most emphatically, no money. Francis encouraged those who followed in his way of life to work for their living if they could (in addition to their ministry to lepers, paupers, and other social outcasts) and beg for subsistence otherwise, but they were to accept neither cash nor landed property. Only in-kind payment needed for immediate survival could be accepted, and the survival horizon was, for Francis and his brothers, the present day. God will provide for tomorrow.

Religious asceticism prior to Francis was just that—asceticism. It was and remains mostly inner-directed, a *quid pro quo* arrangement with God: I'll deny myself worldly comforts if you grant me spiritual enlightenment. Francis' asceticism was outer-directed toward the poor and destitute he found crowding early Italian society. The aim of poverty was to emulate them, for they, the beatitudes of Luke's Gospel tell us, are closer to God and shall therefore inherit the earth (Matthew's later Gospel would refocus attention on the poor *in spirit*, a difference wealthy Christians would exploit for centuries).

The practical consequence of imitating the poor—who were themselves already "imitating" the poor Christ, whether or not they consented to their lot—was to renounce wealth in all of its forms and to avoid fooling oneself about just what forms it takes. It was common in Francis' day for monastic orders (those in which cloistered monks pray and labor throughout the day, as opposed to practicing ministry in the world) to point out that no individual monk owns anything or has access to personal comforts, while the abbey or the order itself could and did own vast expanses of lucrative, productive land, opulent monasteries, and fortunes in gold and other fineries. Personal property was out, but corporate property was permitted and quite abundant. Francis was determined that corporate property would not accumulate among his "Lesser Brothers" any more than personal property would. He thus prohibited the financing and building of permanent houses, schools, libraries, and the like, and discouraged the brothers from inheriting these things from gracious benefactors or by settling down in places that might provide a tempting practical benefit of long-term survival. Francis thus founded one of the first *mendicant* orders—collections of people vowed to religious life but who eschew the stability of sustained residence in one place and who depend upon the goodness of those to whom they minister and others for their day-to-day subsistence. Dominic de Guzman would be doing much the same thing in Spain about that time by establishing his Dominican order, and his fellow Spaniard Ignatius of Loyola similarly adopted this path three centuries later with the Jesuits.

Twenty years prior to my return to the church and a full quarter-century before entering Franciscan life, I had my first encounter with the nascent Green movement in the U.S. via the "Committees of Correspondence" (CoC), an umbrella organization of local cells consisting in like-minded people who found each other by writing to a P.O. Box at Macalester College in Minnesota, if only to learn by post weeks later that a neighbor four doors down also considered herself "green" (such was political organizing before Facebook and Twitter). The group eschewed hierarchical political organization, "leaders," and anything that might smack of the centralized control they were trying to undermine in contemporary society. Many members sought out the organization after being inspired by the recent wins in the German Bundestag by Die Grünen—the most successful (and flamboyant) national Green Party at the time, though the CoC Greens were largely disenchanted with electoral politics in the American context and often disparaged their counterparts in the equally fledgling Green Party USA for its commitment to what these CoC Greens considered outmoded electoral, partisan politics ("neither left nor right but out in front" was the slogan). Despite the foreign inspiration, the "Committees of Correspondence" name was borrowed from a similarly decentralized network of subversives struggling against British tyranny during the American Revolutionary War. At some point, the two ecologically minded organizations—the Green Party USA, a political vehicle with an environmental agenda, and the CoC, a loose band of aging and neophyte hippies, self-styled anarchists, and the stereotypical tree-huggers—merged their mailing lists and became the Green Party we know today in the United States. It's the one that ran Ralph Nader as a presidential candidate twice, despite the fact that he reportedly never in fact became a member of the party.

I became involved with a branch of the CoC in 1985, joining the local chapter along with Ike Balbus—a political theorist and my academic advisor at the time. The Chicago contingent was all of about a dozen people. We often met in the Lincoln Park living room of one member who was in close contact with the national umbrella organization, though it was made abundantly clear he was not "leading" the group or "chairing" our meetings, despite the fact that what he did looked a lot like leading and chairing ("facilitating" was the euphemism we used to describe his getting us through the order of business and scheduling future activities). Meetings ranged from the mildly pleasant and cordial to the insufferable. We typically convened informally around food, making a vegetarian pot luck out of each opportunity to come together. But it soon became apparent that the variety of backgrounds attracted to the early Green movement would not mix well under all circumstances. New Age spiritualists dominated our cell, and the few of us who clung to Marxist or radical anarchist principles were marginalized as relics of the "old paradigm."

What unsettled me most during my participation with the Chicago Greens, however, was the posturing about "voluntary poverty." These profoundly well-to-do people of various ages found their sole environmental strategy in renouncing their class privileges and comforts and persuading those of similar means to do the same. "Humanity"—in reality a heterogeneous organism of classes, genders, and ethnicities each with their own advantages or (mostly) burdens depending upon how you sliced them—was for the local Greens the homogenous enemy of an otherwise unspoiled "Nature." We all, it seemed, had to reduce our impact on the planet, and politics begins at home, with our own lifestyles. I would occasionally remind these earnest inactivists that many of the people inhabiting the earth could in fact stand to *increase* their impact on the planet. They are among the millions whose carbon footprint is negligible largely because they have no sources of manufactured energy in their daily lives, who do not overconsume water because they have at best infrequent access to potable sources, and who waste no food insofar as they can scarcely secure enough to get them through the week. Our space-alien friend might well assume out of naiveté that we all shared in the wealth to similar excesses, and that the responsibility for the ensuing problems was thus equally well-distributed. The self-described "deep-ecologists" I encountered (few used the generic label "environmentalist," given its negative associations with merely "reformist" politics) had no similar excuse; they were simply uninterested in "old-paradigm" bread-and-butter issues like class struggle. Most were content to pronounce their misanthropic generalities, few of which acknowledged that the environmental malaise stemming from their past overconsumption is quite consistent with widespread poverty—as was, I believed, the "voluntary simplicity" to which they aspired.

How could anyone quibble with so noble an ambition as to live lightly on the planet and in solidarity with the poor, unencumbered by the comforts of this world and, for some like Franciscans and other mendicant orders, surrendered to the graces of God? "Well-intentioned but naïve," some might say (many do, of course), but wrong or misguided?

There is something especially peculiar about the contemporary, continued romanticization of poverty by people of faith, and our seeming refusal to acknowledge that it doesn't seem to have done a damn thing about real, involuntary poverty in the world. Secular society has its own incentives for erecting a "poverty industry" that serves to maintain the plight of the poor all the while lamenting it and making gestures to alleviate it. It's all a bit more understandable in a civil context, especially through a Marxist lens, which elaborates quite effectively on the various ways in which the privileges of capitalism insulate themselves from rebuke in part by co-opting the critiques of the system and its excesses and deflecting them via "harmless" channels; it can manufacture a cadre of "the concerned" and "the generous" who reform at the edges while (perhaps unwittingly) keeping the core intact—in this case, those who may relieve some of the discomforts of the poor or alter public perceptions of them without actually challenging the circumstances of their impoverished condition. Marx of course also had concerns about poverty, and he had a good deal of sympathy and esteem for those forced to live lives of drudgery, however little faith he had in the absolutely destitute to rectify their situation (it was the industrial working class on which Marx placed his bets). But even those of us who maintain some fidelity to the Marxist legacy would rightly think Marx a bit insincere if, in the end, he were to admonish us to keep a certain amount of proletarian squalor around so that we could continue to emulate it and find it an ongoing source of character-building for the rest of us who might dabble in it now and again. The socialist revolution, however remote its possibility, and however it might pale in comparison to the Kingdom of God on earth, would eradicate dire want, not attenuate and ennoble it.

In fairness, my complaint is and has been not that voluntary poverty is somehow "wrong" but rather that it is—be it the result of a religious vow or a political statement of lifestyle choices—anachronistic: it is attuned to another era and thus misplaced in ours. It no longer carries the meaning it did for men and women in the middle ages when voluntary religious poverty of the sort St. Francis lived was first conceived and when it couldn't possibly have anything to do with our modern notion of *ecology*.

Prior to the age of industrial capitalism, to lack the necessities of life had a potentially universal air to it, however lamentable the condition may have been. The poor were shunned and disadvantaged to be sure, but because poverty was at least in part a condition given by "natural" resource scarcity, anyone could be subject to it. Yes, there were privileged aristocrats and there were dispossessed peasant classes, but status alone couldn't fully insulate one against the presumed caprices of God: natural catastrophe, famine, etc. The rich would eat first and ultimately the most, yet it was conceivable in difficult times that no one would have enough.

Given its seeming inevitability and its occasional disregard for class discrimination, poverty was something worthy of emulation and sanctification even if one in fact did have access to sufficient means but chose to renounce it for loftier aims. If there is no hope for curing the condition of the downtrodden, to give it all away and stand with those downtrodden signals a certain dignity in indignity, a sense of the graces of God. When Francis would, as the hagiographers tell us, give food to someone worse off—food that he had himself just begged for himself or his brothers—his point was something along the lines of, "If this suffering must exist, make me a part of it, for I will never have comfort as long as one of my brothers or sisters is in deprivation." This is in many ways *the* Christian moral message, and its impact was quite profound in a world where, from all appearances, such suffering by at least part of the population was unavoidable. God has deemed that some must go without adequate sustenance; far be it from me to be among those who flourish and fatten.

Modernity changed all of this, or at least it should have. The natural elements and hazards still cause disruptions, of course, and they may still strike some as acts of a vengeful God, but we are not entirely at their mercy. Though the world's population is substantially larger than it was at the turn of the first millennium CE, we now have the capacity to feed it, house it, and keep it generally free from basic harm. We have the technology and the arable land needed to sustain the global population for the short term, and a glimpse at how to do so in ecologically sound ways for the long term. What we lack is the political will. Poverty has ceased to be strictly a function of natural conditions handed to us by God, and is now a matter of political and economic sclerosis. We quibble over the presumed causes—indolence, dependency, backward institutions, regulatory stifling of markets, the ravages of capitalism, etc.—yet the identification of symptoms is the same: we can produce enough to feed us all, but we squander that product in such a way that, still, some people are overfed, wasteful, and hazardous to the planet while, still, others starve.

To mimic poverty under these circumstances, even if by mimicking it we aim to alleviate somewhat the condition of the genuinely poor, carries a very different meaning from the prescriptions of St. Francis in terms of our relation to the poor, to the natural world, and ultimately to God. Francis gave us the first form of voluntary poverty as *solidarity* with the actually poor, the first renunciation of material wealth that said "I can't fix your condition, but I can choose to bear it with you." The problem with such displays of solidarity in our current era is the first clause: while we can't literally "fix" poverty or our ecological state of our own accord—since, as we've seen, an equitable and environmentally sound society amounts to a *public good*—we can in fact contribute to their solutions, however small that contribution might be and however intractable the problem might seem (we are, in fact, as I've argued in this book, *obliged to* in a very real sense). Mere "solidarity" with the poor and the dispossessed remains a potentially laudable aim, but if it does nothing to improve the condition of those who are poor by no choice of their own or the planet that we have jeopardized from under their feet, such solidarity is feeble at best. If I'm lacking in the basic necessities of life while others can squander their excess, it means little to me that you elect to surrender your claim on that excess in order to live like me—especially if your own spiritual salvation is a large part of your motivation.

The voluntary poverty of the twenty-first century is thus the performative poverty of the select few in a world where one can pursue personal satisfaction or spiritual enlightenment without regard to its political effect upon others in changing their lives. This is just as well, since it would in fact be impossible for masses of like-minded people to follow this lead and to live with the same personal ends of "simplicity" or "perfection" in mind and yet co-exist with a planet of more than seven billion people. To the extent that voluntary poverty in the way St. Francis lived it can make sense in our age, it can only make sense for a very small subset of the human population. We can't all live in "voluntary poverty" of this sort, nor would any of us likely find livable or desirable a world in which we all tried. Such a self-conscious "poverty" remains wholly dependent on the productive capacity that privileges the few, exploits the many, and despoils the earth.

We vowed religious are well aware of this fact, of course, and have long accepted it as the uniqueness of religious life. All Christians, we have believed for centuries, are called to live the "Gospel life"—that is, to love one another and God as we love ourselves and as God loves us, to forgive one's enemies, to observe the moral codes of the Hebrew Decalogue, etc. These are the moral precepts that go with being a Christian; they are required of all of us as a matter of duty. Those of us who sense a call to life in a religious congregation are further called to live the evangelical counsels: celibate chastity, obedience, and poverty, or what we loosely call "the vows." The counsels are indeed binding upon a particular individual once he or she has made a solemn profession to live them, but they are not required of all Christians. As counsels, they are considered morally commendable and good if indeed one senses the call to vowed religious life, but one is otherwise not morally culpable or "deficient" in opting not to live them. In the technical vernacular of moral philosophy seen earlier in this text, they are *supererogatory* but not universally *obligatory*: they are not matters of duty, but they go quite generously above and beyond duty's imperatives if one does elect to pursue them. In short, they were never intended for mass adoption as exemplary practices suitable for *everyone*.

It was, in fact, the supererogatory nature of Francis of Assisi's strident understanding of "evangelical poverty" that contributed to our modern moral distinction between the duties binding upon us all and those actions which are commendable if aspired to and achieved but not demanded of us. By the mid-thirteenth century, the "schoolmasters" at the University of Paris (professors and "secular" priests answering to local bishops rather than superiors in the religious orders) had had enough of the Franciscan and Dominican prattling on about their "poverty." They think, the schoolmasters grumbled among each other, that they're better than us because they renounce the luxuries of the church. Gerrard of Abbeville penned a scathing sermon that vilified the Franciscan clerics and professors at the University. In equating their austere lifestyle with the proper Christian life, he complained, they effectively pronounced most of Christendom *sinners*, insofar as no one else lives up to such a standard of poverty. Bonaventure of Bagnoregio, then Minister General of the Franciscan order, wrote *Defense of the Mendicants* in reply to Gerrard, suggesting among other things that such a charge was nonsense. This life is not asked of everyone; it is required only of those who commit to it, who elect to strive for "evangelical perfection." We can't expect all of us to be perfect in this way. This was among the earliest uses of a distinction between moral *duties* and morally commendable (but not obligatory), *supererogatory* acts.

Bonaventure's point declawed Gerrard and the schoolmasters somewhat, but it can't be a sufficient response in our world now. Voluntary poverty—be it an act of solidarity with the actually dispossessed or be it an attempt to live one's ideal of an ecologically sound future in the here and now—cannot be the distinguishing feature of the select few who opt to go the extra mile on behalf of their values, a feeble "witness" that the faithful and anyone else can applaud from a comfortable distance. It must be a practical example of living that those with or without faith, motivated toward similar ends by different reasons, can themselves live on behalf of the genuinely dispossessed and on behalf of a troubled planet—until the point that both conditions are eliminated and no one's "poverty" is appreciably different from anyone's wealth. It must be, in other words, political and not purely spiritual in its ends. And it must be understood not as a lifestyle option but obligatory, a duty required of us all.

Philosophy is not made better (or worse) by having something as grand as a "personal journey" to tack on to the end of one's conclusions, but I am willing to concede that this book has had a strange evolution. The principal arguments laid out in the preceding chapters were formulated at a time in my life when I was a long-lapsed Catholic and a professed agnostic. The notion that philosophy alone could be used to discern and highlight the demands of morality that fall upon each of us seemed perfectly natural to me. It still does; this much hasn't changed. I still believe it is legitimate and necessary to say to the environmental foot-dragger and the libertarian that the moral weight of ecology binds us all, that we all have identifiable duties to each other in providing for a healthy natural world, and that the relative force of those duties cannot be calculated much less "bought off" by market means alone. And I still believe it legitimate and necessary to say to the self-identified environmentalist content with a few minor "voluntary" lifestyle adaptations that approximate what he or she imagines the "simple" life of poverty to be that, no, this will require much more from you than a tweak here and there to your consumer habits, and that you must fully render your life a challenge to this system that degrades the planet on behalf of the few and at the expense of the many. The call to socialism of one sort or another cannot be dodged by good intentions, however much they may be motivated by sincere environmental concerns.

The fact that my life is now formally vowed to the tradition of "evangelical poverty" as inspired by St. Francis of Assisi eight centuries ago has altered somewhat my perspective, however. I must now speak not only as a political philosopher and teacher but also as a minister. Academic pedagogy is no longer a sufficient location from which to express these ideas; they must be *pastoral* as well. Fortunately, this personal evolution has little impact on the substantive issues at hand. I can still "preach" that matters of ecology are inseparable from matters of economics, and that how we treat the planet is intimately tied up with how and whether we share this planet's abundant resources with each other. And even more fortunately by way of impeccable timing, this "preaching" reaches print mere months after Pope Francis (who, of course, though himself a Jesuit, took his name from my congregation's patron) published his encyclical *Laudato Si*, named from the refrain of Francis of Assisi's *Canticle*, and pleading much the same connection between ecological healing and poverty relief that I've long held.

Permit me, then, to close this text by restating what I wish the reader to take from it in light of my pastoral commitments and in light of the Holy Father's challenge to the status quo of poverty and ecological negligence in this world.

Environmentalism without an equally abiding passion for an equitable distribution of the planet's wealth and resources is pointless. Few environmentalists or anyone else would admit, when pressed, that their vision of a sustainable planet lacks a distributive reference point, a theory of how the planet's wealth should be equitably shared as well as prudently cultivated.

The track record of the environmental movement doesn't corroborate any long-standing dual concern, however. At best, we hear the (perfectly justifiable) claim that excessive environmental damage is to be condemned in part because of its disproportionate, adverse effects on the poor. Only occasionally do we hear from the environmental movement the hypothesis that perhaps the poor *are in fact poor in the first place* as a consequence of the same processes and decisions that imperil the planet's ecology. I have not argued in this book on behalf of the redistribution of wealth or its relationship to the "moral weight of ecology," but I would be undermining my own point if I didn't emphasize that a future society that has managed to contain the worst of our ecological excesses without giving the planet's dispossessed the means to a viable existence is morally no more defensible than the world we have now, if indeed it is conceivable at all.

Poverty is not a permanent condition of humanity to be forever emulated by some and periodically relieved through charitable works but a terminal blight that is to be eradicated once and for all. Again, most people might pay cheerful lip service to this proposition all the while believing, deep down, that we're stuck indefinitely with extreme want in some substantial portion of the population. Christians who tend to cherry-pick their scriptural passages in order to lend the most credence to their un-Christian predispositions will follow this with, "Christ himself said poverty will always be with us." Secular society now chants a similar mantra: the poor you will always have with you because the Enlightenment project of bettering the lot of humanity has been pronounced immature if not "oppressive" by the fashions of postmodern cynicism. As if to raise a middle finger to that cynicism, every year thousands of people who might be morally justified in just tending to their own needs and local responsibilities put it all aside and devote their efforts to ministering to the immediate needs of the impoverished or to the viability of the ecosphere. We do not under any circumstances want to discourage this or diminish those efforts. But we must also remind them and ourselves that these efforts ought to be about the poor or about the planet and not about us. If such efforts, however invaluable, are done strictly for purposes of personal redemption or the expiations of one's own sins (or, as I fear in some cases, to round out a résumé), or if they are accompanied by the belief that the need and duty to eradicate poverty and environmental degradation ends with the close of one's stint with the Peace Corps or the Franciscan Mission Service or the World Wildlife Fund, they are misguided. For every kid that exits high school or college with the wistful desire to "work with the poor," we now need ten more with a passion for putting that kid—bless his or her heart—out of business. The poor will indeed be with us, but only for as long as we believe we can somehow dispatch our duties to them while the condition of poverty remains.

Mutual sustainability and flourishing, not material self-deprivation, is the proper goal of the thoughtful life. The notion of voluntary poverty, religious or otherwise, is as I've argued here no longer meaningful in the sense it once was. It is far too easy to live it in the abstract, without regard to actual poverty in the world or how one is implicated in perpetuating it. If a "lifestyle" choice of any substance can be made that would demonstrate solidarity with the actually poor, it more likely would involve making one's life materially compatible with that of everyone else on the planet. Pitching one's aims negatively in terms of what one imagines "poverty" to be and aspiring to parrot it misses the point. The aim is not to scale down one's comforts for its own sake, much less to "out-poverty" others in a competition over self-imposed austerity; rather, the aim is to enjoy the fruits of God's green earth as much as is consistent, in principle at least, with the same enjoyment by the world's population. This, arguably, is what environmentalism is all about, and it is the point of the preceding chapters in a nutshell. The environmental movement has been quite rightly describing this as a form of *sustainability* for years, and I see no reason to reinvent the wheel or rebrand it. Yet it must be remembered that, when we do scale back our personal impact on the plant, we are doing so not merely for the sake of that planet's integrity but so that others might tread with a bit *more* weight than they have been permitted to tread until now. The balance may be a difficult one to identify, but as I've argued throughout this text, we have the tools of economics, ecology, and moral philosophy to help locate it.

Duty obliges, but only in a formal sense; we must be somehow predisposed to act upon duties that bind us for them to be meaningful. To the ears of one attuned to religious faith, and perhaps even some without a particular faith, much of what I've argued throughout this book can be at best only half the story. The "moral weight of ecology" may well demand of us particular changes in our lives in the interest of a more sustainable world, and it may undergird the legal justification for social rules that redirect the way we use resources, but duty itself does not *motivate*. Duty gives us a *reason* for action and perhaps a justification for the threat of *coercion* (two very different things, as we saw), but it does not by itself impel action. This isn't to suggest that we have to *want* to adhere to the duties that morally bind us for them to be meaningful (they wouldn't really be *duties* if they required our consent). The point, rather, is that discerning which moral imperatives are incumbent upon us by itself has no necessary practical effect in the world.

Faith, I've come to believe, is less about the interior life of prayer and contemplation than it is about the *exterior* life—the appreciation of habits, the little daily rituals, and what Wittgenstein described as the forms and practices of life that give meaning to our language and our social interactions.

To contrast these small, seemingly inconsequential practices with the larger structures that shape us, as I had done for many years, inevitably ends up making them appear smaller and even more inconsequential. This, in a sense, is the source of the old Marxist impatience with the utopian socialists bent on changing the world one intentional commune at a time by re-knitting the micro-fabric of human life within them—an emphasis on the "demand" side of the equation alone rather than the "supply side" of the forces and relations of production. You can't change society, Marx insisted, without changing the social structures by which we materially reproduce our existence; give me five minutes alone with every peasant in the countryside, the Russian anarchist Mikhail Bakunin reportedly responded: you'll have your revolution.

But comparing the habits of daily life with "base and superstructural" phenomena is not the only or even the best comparison. Better to compare them to inaction, indifference, and complacency—those tendencies that the modern environmental movement has had to contend with for a few generations now. Plotting social transformation at the structural level alone, be it by the overthrow of a tyrant or the seizing of the means of production, cleaves one's time between the political and the mundane; even the revolutionary must pause to cook dinner and change the baby's diapers, and these may count for him or her as distractions from the *real* cause. The architecture of social transformation that the ecology movement draws up, at its politically most effective, offers no such down time. We are permanently on the clock, and every activity is under scrutiny insofar as it either contributes to a more sustainable world or it doesn't. Yes, duties bind us, but until we can internalize those moral requirements in a healthy way, guided by mutual love of person, planet, and, for those who chose to find meaning in a higher source, the God that creates both, duties are inconsequential. The "moral weight of ecology," as with any weight, is only measurable in an atmosphere; things have no mass in a vacuum without air, gravity, etc. We must provide that "atmosphere" ourselves, and this is precisely what our ministry to each other as human beings does.

Toward the end of *Laudato Si*, Pope Francis raises a similar caution regarding merely "educating" people about the fragility of the natural environment and erecting legal structures aimed at reinforcing what is learned. Such an education is indispensable, of course, if anything is to be done.

> Yet this education, aimed at creating an 'ecological citizenship,' is at times limited to providing information, and fails to instill good habits. The existence of laws and regulations is insufficient in the long run to curb bad conduct, even when effective means of enforcement are present. If the laws are to bring about significant, long-lasting effects, the majority of the members of society must be adequately motivated to accept them, and personally transformed to respond.

> Only by cultivating sound virtues will people be able to make a selfless ecological commitment. A person who could afford to spend and consume more but regularly uses less heating and wears warmer clothes, shows the kind of convictions and attitudes which help to protect the environment. There is a nobility in the duty to care for creation through little daily actions . . . [they] reflect a generous and worthy creativity which brings out the best in human beings. (Francis I, 2015:#211)

Conservative pundits—secular and religious—have taken to describing the Holy Father as a "radical," a "liberation theologian," even a "Marxist," especially after the publication of *Laudato Si*. Liberal pundits in turn suggest that he's merely ratifying the message of the mainstream environmental movement and admonishing us about thermostats and grocery bags. If only either could grasp the revolution Pope Francis is actually calling for.

We who inhabit the cultures that have fostered consumer excess and the despoliation of the earth do not take kindly to being told we must change the way we live. When circumstances require it and a collective will executes such a demand, we implement it reluctantly through the polite institutions of statute, regulation, and taxation intended to discourage certain actions and encourage others. Those who champion our liberties quite justifiably look upon these institutions with suspicion; the imposition of our democratic will upon persons who may not see things our way is never a trivial matter. But impose we must, and, occasionally, we must do so with the coercive threat of force embodied in government when markets and their incentives fail. In the sterile dystopias of the libertarian imagination, there is never remorse or lament for this fact. In the real world of sentient human beings, however, to ask one to alter the very way she goes about living on this delicate little planet is never strictly one-sided: it takes a toll on the one who asks, whether that individual realizes it or not. I've suggested in these pages that, yes, you will in your lifetime very likely be required to change the way you inhabit this world if you wish to see it thrive—and, just as importantly, even if you don't particularly care. But the process of anticipating and justifying this demand upon you has changed me as well. Such are the ways of God.

Bibliography

Ackerman, Frank. "Cost-Benefit Analysis of Climate Change: Where it Goes Wrong," in David M. Driesen (ed.), *Economic Thought and U.S. Climate Change Policy*, 2010: 61–81.

Adamowicz, Vic. "Alternative Valuation Techniques: A Comparison and Movement to a Synthesis," in R. G. Willis and J. T. Corkindale (eds.), *Environmental Valuation: New Perspectives*, 1995: 144–159.

Anderson, Terry and Leal, Donald. *Free Market Environmentalism*. San Francisco: Pacific Research Institute for Public Policy, 1991.

Anderson, Terry and Leal, Donald. "Free Market Versus Political Environmentalism," *Harvard Journal of Law and Public Policy* 15,2 (Spring 1992): 97–310.

Arneson, Richard J. "The Principle of Fairness and Free-Rider Problems," *Ethics* 92,4 (July 1982): 616–633.

Arrow, K., Solow, R., Protney, P., Learner, E., Radner, R. and Schuman, H. "Report of the NOAA Panel on Contingent Valuation," U.S. Federal Register 58,10 (1993): 4601–4614.

Baldwin, Richard. "Does Sustainability Require Growth?" in Ian Goldin and L. Alan Winters (eds.), *The Economics of Sustainable Development*, 1995: 51–78.

Barbier, Edward B. *Economics, Natural Resource Scarcity, and Development*. London: Earthscan Publications Limited, 1989.

Barbier, Edward B. (ed.). *Economics and Ecology: New Frontiers in Sustainable Development*. London: Chapman and Hall, 1993.

Baudrillard, Jean. *For a Critique of the Political Economy of the Sign*. Translated by Charles Levin. St. Louis: Telos Press, 1981.

Becker, Gary. *A Treatise on the Family*. Cambridge, MA: Harvard University Press, 1981.

Beckerman, Wilfred. *In Defence of Economic Growth*. London: Jonathan Cape, 1974.

Bell, Nora. "Nozick and the Principle of Fairness," *Social Theory and Practice* 5,1 (Fall 1978): 65–73.

Blumm, Michael C. "The Fallacies of Free Market Environmentalism," *Harvard Journal of Law & Public Policy* 15,2 (Spring 1992): 370–389.

Brown, Gardner Jr. "The Viewing Value of Elephants," in Edward Barbier (ed.), *Economics and Ecology*, 1993: 146–155.

Brunet, Edward. "Debunking Wholesale Private Enforcement of Environmental Rights," *Harvard Journal of Law & Public Policy* 15,2 (Spring 1992): 310–323.

Chandler, William U. "Banishing Tobacco," in Linda Starke (ed.), *State of the World 1986*, 1986: 139–158.

Clarke, Edward H. "Experimenting with Public Goods Pricing: A Comment," *Public Choice* 23 (Summer 1975): 35–48.

Clarke, Edward H. *Demand Revelation and the Provision of Public Goods.* Cambridge, MA: Ballinger, 1980.

Coase, R. H. "The Problem of Social Cost," *The Journal of Law and Economics* 3 (October 1960): 1–44.

Cocking, Dean and Oakley, Justin. "Indirect Consequentialism, Friendship, and the Problem of Alienation," *Ethics* 106, 1 (October 1995): 86–111.

Costanza, Robert (ed.). *Environmental Economics.* New York: Columbia University Press, 1991.

Cranor, Carl. "Collective and Individual Duties to Address Global Warming," in David M. Driesen (ed.), *Economic Thought and U.S. Climate Change Policy*, 2010: 153–169.

Cullity, Garrett. "Moral Free Riding," *Philosophy and Public Affairs* 24,1 (Winter 1995): 3–34.

Daley, Herman E. (ed.). *Economics, Ecology, Ethics.* San Francisco: W.H. Freeman and Co., 1973.

Daley, Herman E. and Cobb, John B. *For the Common Good.* Boston: Beacon Press, 1989.

Daly, Herman E. *Steady-State Economics.* Second edition. Washington, D.C: Island Press, 1991.

De Jasay, Anthony. *Social Contract, Free Ride.* Oxford: Clarendon Press, 1989.

Demsetz, Harold. "Toward a Theory of Property Rights," *American Economic Review* 57 (May 1967): 347–359.

Devall, Bill. "The Deep Ecology Movement," *Natural Resources Journal* 20,1 (April 1980): 299–322.

Devall, Bill and Sessions, George. "The Development of Natural Resources and the Integrity of Nature," *Environmental Ethics* 6,4 (Winter 1984): 293–322.

Devall, Bill and Sessions, George. *Deep Ecology: Living as if Nature Mattered.* Salt Lake City: Gibbs M. Smith, 1985.

Diamond, Peter A. and Hausman, Jerry A. "On Contingent Valuation Measurement and Nonuse Values," in Jerry A. Hausman (ed.), *Contingent Valuation: A Critical Assessment*, 1993: 3–31.

Dore, Mohammed H. I. "The Problem of Valuation in Neoclassical Environmental Economics," *Environmental Ethics* 18,1 (Spring 1996): 64–70.

Driesen, David M. (ed.). *Economic Thought and U.S. Climate Change Policy.* Cambridge, MA: The MIT Press, 2010.

Edwards, Steven. "In Defense of Environmental Economics," *Environmental Ethics*, 10,1 (Spring 1987): 73–85.

El Serafy, Salah. "The Environment as Capital," in Robert Costanza (ed.) *Environmental Economics*, 1991: 168–175.

Elster, Jon. *Logic and Society*. Chichester: John Wiley and Sons, 1978.

Elster, Jon. *Making Sense of Marx*. Cambridge: Cambridge University Press, 1985.

Epstein, Richard. *Takings, Private Property, and the Power of Eminent Domain*. Cambridge, MA. Harvard University Press, 1985.

FAO. *Livestock's Long Shadow: Environmental Issues and Options*. Report of the Food and Agriculture Organization of the United Nations, 2006. Accessed June 24, 2015. http://www.fao.org/docrep/010/a0701e/a0701e00.HTM.

Feller, William. *An Introduction to Probability Theory and Its Applications*. New York: John Wiley & Sons, 1968.

Flavin, Christopher and Engelmen, Robert. "The Perfect Storm," in Linda Starke (ed.), *State of the World 2009*, 2009: 5–12.

Foster, John (ed). *Valuing Nature?: Economics, Ethics, and Environment*. New York: Routledge, 1997.

Foster, John Bellamy. "The Limits of Environmentalism Without Class: Lessons From the Ancient Forest Struggle in the Pacific Northwest," *Capitalism, Nature, Socialism* 4,1 (March 1993): 11–41.

Francis I. *On Care for Our Common Home—Laudato Si ("Praised Be")*. May 24, 2015. Accessed June 24, 2015. http://w2.vatican.va/content/francesco/en/encyclicals/documents/papa-francesco_20150524_enciclica-laudato-si.html.

Funk, William. "Free Market Environmentalism: Wonder Drug or Snake Oil?" *Harvard Journal of Law and Public Policy* 15,2 (Spring 1992): 511–516.

Gardiner, Stephen M. *A Perfect Moral Storm: The Ethical Tragedy of Climate Change*. Reprint edition. New York: Oxford University Press, 2013.

Gillroy, John Martin (ed.). *Environmental Risk, Environmental Values, and Political Choices*. Boulder: Westview Press, 1993.

The Global 2000 Report to the President of the U.S.: Entering the 21st Century; Volume I: The Summary Report. New York: Pergamon Press, 1980.

Goldin, Ian and Winters, L. Alan (eds.). *The Economics of Sustainable Development*. Cambridge: Cambridge University Press, 1995.

Gowdy, John M. and Olsen, Peg R. "Further Problems with Neoclassical Environmental Economics," *Environmental Ethics* 16,2 (Summer 1994): 161–171.

Grossman, Gene M. "Pollution and Growth: What Do We Know?" in Ian Goldin and L. Alan Winters (eds.), *The Economics of Sustainable Development*, 1995: 19–46.

Grundman, Reiner. *Marxism and Ecology*. Oxford: Clarendon Press, 1991.

Hahnel, Robin. *Green Economics: Confronting the Ecological Crisis*. Armonk, NY: Routledge, 2011.

Hampton, Jean. "Free Rider Problems in the Production of Collective Goods," *Economics and Philosophy* 3 (1987): 245–273.

Hanemann, W. Michael. "Contingent Valuation and Economics," in R. G. Willis and J. T. Corkindale (eds.), *Environmental Valuation: New Perspectives*, 1995: 80–117.

Hanly, Ken. "The Problem of Social Cost: Coase's Economics Versus Ethics," *Journal of Applied Philosophy* 9,1 (1992): 77–83.

Hansel, Saul. "Wave of Mergers is Transforming Banking," *The New York Times*, August 21, 1995, sec. A: 1, 9.

Hardin, Garrett and Baden, John (eds.). *Managing the Commons*. San Francisco: W.H. Freeman and Company, 1977.

Hardin, Garrett. "The Tragedy of the Commons," in Garrett Hardin and John Baden (eds.), *Managing the Commons*, 1977:16–30, reprinted from *Science*, 162 (1968): 1243–1248.

Hardin, Russell. *Collective Action*. Baltimore: Johns Hopkins University Press, 1982.

Hare, W. L. "A Safe Landing for the Climate," in Linda Starke (ed.), *State of the World, 2009*, 2009: 13–29.

Hargreaves Heap, Shaun and Varoufakis, Yanis. *Game Theory*. London: Routledge, 1995.

Harrison, Glenn W. "Valuing Public Goods with the Contingent Valuation Method: A Critique of Kahneman and Knetsch," *Journal of Environmental Economics and Management* 23 (November 1992): 248–257.

Hart, H. L. A. "Are There Any Natural Rights," *Philosophical Review* 64 (1955): 175–191.

Hausman, Jerry A. (ed.). *Contingent Valuation: A Critical Assessment*. Amsterdam: North-Holland, 1993.

Hawken, Paul. *The Ecology of Commerce*. New York: Harper Collins, 1993.

Haworth, Alan. *Anti-Libertarianism: Markets, Philosophy, and Myth*. London: Routledge, 1994.

Hays, Sam. "Emissions Trading Mythology," *Environmental Forum* 12,1 (January–February 1995): 15–20.

Head, J. G. "Public Goods and Public Policy," *Public Finance* 17 (1962): 198–219.

Heyne, Paul. "Economics, Ethics, and Ecology," in Roger E. Meiners and Bruce Yandle (eds.), *Taking the Environment Seriously*, 1993: 25–49.

Hicks, J. G. *Value and Capital*. 2nd ed. Oxford University Press, 1939.

Hoevenagel, Ruud. "An Assessment of Contingent Valuation Surveys," in Ståle Navrud (ed.), *Pricing the European Environment*, 1992: 177–194.

Holland, Allan. "The Assumptions of Cost-Benefit Analysis: A Philosopher's View," in R. G. Willis and J. T. Corkindale (eds.), *Environmental Valuation: New Perspectives*, 1995: 21–38.

Huffman, James L. "Protecting the Environment from Orthodox Environmentalism," *Harvard Journal of Law and Public Policy* 15,2 (Spring 1992): 349–370.

Hummel, Jeffrey Rogers. "National Goods Versus Public Goods: Defense, Disarmament, and Free Riders," in Murray N. Rothbard (ed.), *The Review of Austrian Economics*, Volume Four. Boston: Kluwer Academic Publishers, 1990: 88–122.

IPCC. *Climate Change 2000: Mitigation*. Contribution of Working Group III to the Third Assessment Report of the Intergovernmental Panel on Climate Change. Accessed June 24, 2015. http://www.ipcc.ch/ipccreports/tar/wg3/index.php?idp=48.

Jacobs, Michael. "The Limits to Neoclassicism: Towards an Institutional Environmental Economics," in Michael Redclift and Ted Benton (eds.), *Social Theory and the Global Environment*, 1994: 67–91.

Johansson, Per-Olov. *The Economic Theory and Measurement of Environmental Benefits*. Cambridge University Press, 1987.

Kagan, Shelly. "The Additive Fallacy," *Ethics* 99,1 (October 1988): 5–31.

Kahneman, Daniel and Jack L. Knetsch. "Valuing Public Goods: The Purchase of Moral Satisfaction," *Journal of Environmental Economics and Management*, 22 (January 1992a): 57–70.

Kahneman, Daniel and Jack L. Knetsch. "Contingent Valuation and the Value of Public Goods: Reply," *Journal of Environmental Economics and Management*, 22 (January 1992b): 90–94.

Keats, Russell. "Values and Preferences in Neo-Classical Environmental Economics," in John Foster (ed.), *Valuing Nature?* 1997: 32–47.

Klein, Naomi. *This Changes Everything: Capitalism vs. The Climate*. New York: Simon & Schuster, 2014.

Klosko, George. *The Principle of Fairness and Political Obligation*. Lanham, MD: Rowman & Littlefield, 1992.

Kneese, Allen V. *Measuring the Benefits of Clean Air and Water*. Washington D.C.: Resources for the Future, 1984.

Knetsch, Jack L. "Environmental Policy Implications of Disparities Between Willingness to Pay and Compensation Demanded Measures of Values," *Journal of Environmental Economics and Management* 18 (1990): 227–237.

Krier, James E. "The Tragedy of the Commons, Part Two," *Harvard journal of Law & Public Policy* 15,2 (Spring 1992): 324–347.

Light, Andrew and Avner De-Shalit (eds.). *Moral and Political Reasoning in Environmental Practice*. Cambridge, MA: The MIT Press, 2003.

Lindberg, Erik. "Six Myths About Climate Change That Liberals Rarely Question." *Transitions* Milwaukee, November 24, 2014. Accessed June 24, 2015. http://transitionmilwaukee.org/profiles/blogs/six-myths-about-climate-change-that-liberals-rarely-question.

Lomborg, Bjørn. *The Skeptical Environmentalist: Measuring the Real State of the World*. 1st edition. Cambridge; New York: Cambridge University Press, 2001.

Lowe, Scott C. "Free Riders, 'Fair Share,' and the Principle of Fair Play," *Public Affairs Quarterly* 10,1 (January 1996): 49–62.

Macaulay, Hugh H. and Bruce Yandle. *Environmental Use and the Marke*t. Lexington, MA: Lexington Books, 1977.

Machan, Tibor (ed.). *The Commons: Its Tragedies and Other Follies*. Stanford, CA: Hoover Institution Press, 2001.

Margolis, Julius. "A Comment on the Pure Theory or Public Expenditure," *Review of Economics and Statistics* 37 (1955): 347–349.

McCarney, Joseph. "Elster, Marx, and Method," *Analyzing Marxism* (*Canadian Journal of Philosophy* Supplementary Volume 15). Calgary: University of Calgary Press, 1989: 135–161.

Meiners, Roger E. and Yandle, Bruce (eds.). *Taking the Environment Seriously*. Lanham, MD: Rowman & Littlefield Publishers, 1993.

Merchant, Carolyn. *Radical Ecology: The Search for a Livable World*. New York: Routledge, 1992.

Milgrom, Paul. "Is Sympathy an Economic Value? Philosophy, Economics, and the Contingent Valuation Method," in Hausman (ed.), *Contingent Valuation: A Critical Assessment*, 1993: 417–435.

Mishan, E. J. "Reflections on Recent Developments in the Concept of External Effects," *Canadian Journal of Economics and Political Science* 31,1 (February 1965): 3–34.

Mishan, E. J. "The Postwar Literature on Externalities: An Interpretive Essay," *Journal of Economic Literature* 9,1 (March 1971): 1–28.

Mishan, E. J. *What Political Economy is All About*. New York: Cambridge University Press, 1982.

Mitchell, Robert Cameron and Carson, Richard T. *Using Surveys to Value Public Goods: The Contingent Valuation Method*. Washington, D.C.: Resources for the Future, 1989.

Mulberg, Jon. *Social Limits to Economic Theory*. London: Routledge, 1995.

Murphy, Liam B. "The Demands of Beneficence," *Philosophy and Public Affairs* 22,4 (Fall 1993): 267–292.

Naess, Arne. "The Shallow and the Deep, Long-Range Ecology Movement," *Inquiry* 16 (1973): 95–100.

Naess, Arne. "A Defence of the Deep Ecology Movement," *Environmental Ethics* 6 (Fall 1984): 265–270.

Naess, Arne. "The Deep Ecological Movement: Some Philosophical Aspects," *Philosophical Inquiry* 8,1–2 (1986): 10–29.

Nordhaus, William D. *A Question of Balance: Weighing the Options on Global Warming Policies*. New Haven: Yale University Press, 2008.

Nove, Alec. *The Economics of Feasible Socialism*. London: George Allen & Unwin, 1983.

Nozick, Robert. *Anarchy, State, and Utopia*. New York: Basic Books, 1974.

Olson, Mancur. *The Logic of Collective Action*. Cambridge: Harvard University Press, 1965.

O'Neill, John. *Ecology, Policy, and Politics*. London: Routledge, 1993.

O'Neill, John. "Value Pluralism, Incommensurability, and Institutions," in John Foster (ed.), *Valuing Nature?* 1997: 75–88.

The Onion. "'How Bad For The Environment Can Throwing Away One Plastic Bottle Be?' 30 Million People Wonder," January 19, 2010. Accessed July 10, 2015. http://www.theonion.com/article/how-bad-for-the-environment-can-throwing-away-one--2892.

The Onion. "Decision To Circle Parking Lot Produces Carbon Emission That Finally Does It," October 8, 2014. Accessed July 10, 2015. http://www.theonion.com/article/decision-to-circle-parking-lot-produces-carbon-emi-37131.

Ostrom, Elinor. *Governing the Commons*. Cambridge: Cambridge University Press, 1990.

Otsuka, Michael. "The Paradox of Group Beneficence," *Philosophy and Public Affairs*, 20,2 (Spring 1991): 132–149.

Parfit, Derek. *Reasons and Persons*. Oxford: Clarendon Press, 1984.

Perrings, Charles. "Ecological and Economic Values," in R. G. Willis and J. T. Corkindale (eds.), *Environmental Valuation: New Perspectives*, 1995: 56–66.

Pigou, A. C. *The Economics of Welfare*. London: MacMillan and Co., 1950.

Plantico, Reuben C. "A Property Rights Strategy for Protecting the Environment: A Comment on Stroup and Goodman," *Harvard Journal of Law and Public Policy* 15,2 (Spring 1992): 427–454.

Porrit, Jonathon. *Seeing Green*. New York: Basil Blackwell, 1985.

Rawls, John. *A Theory of Justice*. Cambridge, MA: Harvard University Press, 1971.

Rothbard, Murray. "Mises and the Role of the Economist in Public Policy," in Jeffrey M. Herbener (ed.), *The Meaning of Ludwig von Mises*, 1993: 193–208.

Sagoff, Mark. "Some Problems with Environmental Economics," *Environmental Ethics* 10,1 (Spring 1988a): 55–74.

Sagoff, Mark. *The Economy of the Earth*. Cambridge: Cambridge University Press, 1988b.

Sagoff, Mark. *Price, Principle, and the Environment.* Cambridge: Cambridge University Press, 2004.

Sahlins, Marshall. *Culture and Practical Reason.* University of Chicago Press, 1976.

Samuelson, Paul. "The Pure Theory of Public Expenditure," *Review of Economics and Statistics* 36 (1954): 387–389.

Samuelson, Paul. "Diagrammatic Exposition of a Theory of Public Expenditure," *Review of Economics and Statistics* 37 (1955): 350–356.

Samuelson, Paul. "Aspects of Public Expenditure Theories," *Review of Economics and Statistics* 40 (1958): 332–338.

Scherer, Donald (ed.), *Upstream/Downstream: Issues in Environmental Ethics*. Philadelphia: Temple University Press, 1990.

Scherr, Bruce A. and Babb, Emerson M. "Pricing Public Goods: An Experiment with Two Proposed Pricing Systems," *Public Choice* 23 (Summer 1975): 35–48.

Schmidtz, David. *The Limits of Government.* Boulder, CA: Westview Press, 1991.

Schmidtz, David. *Rational Choice and Moral Agency*. Princeton, NJ: Princeton University Press, 1995.

Schmidtz, David and Robert Goodin. *Social Welfare and Individual Responsibility: For and Against*. Cambridge: Cambridge University Press, 1998.

Schor, Juliet B. *Plenitude: The New Economics of True Wealth*. New York, NY: The Penguin Press, 2010.

Sen, Amartya K. "Rational Fools: A Critique of the Behavioral Foundations of Economic Theory, *Philosophy and Public Affairs* 6 (1977): 317–44.

Shrader-Frechette, Kristin. *Risk Analysis and Scientific Method.* Doldrecht: D. Reidel, 1985.

Simmel, Georg. *The Sociology of Georg Simmel.* Translated and edited by Kurt H. Wolff. New York: Free Press, 1950.

Simmons, A. John. *Moral Principles and Political Obligation.* Princeton, NJ: Princeton University Press, 1979.

Simon, Julian L. *The Ultimate Resource*. Princeton, NJ: Princeton University Press, 1981.

Singer, Peter. "Famine, Affluence, and Morality." *Philosophy and Public Affairs* 1,3 (1972): 229–43.

Slovic, Paul. 1991. "Beyond Numbers: A Broader Perspective on Risk Perception and Risk Communication," in Deborah G. Mayo and Rachelle D. Hollander (eds.), *Acceptable Evidence*, 1991: 48–65.

Smith, Adam. *An Inquiry into the Nature and Causes of the Wealth of Nations*. The University of Chicago Press, 1976 (1776).

Smith, Robert J. "Resolving the Tragedy of the Commons by Creating Private Property Rights in Wildlife," *Cato Journal* 1,2 (Fall 1981): 439–468.

Smith, V. Kerry. "Arbitrary Values, Good Causes, and Premature Verdicts," *Journal of Environmental Economics and Management* 22 (January 1992): 71–89.

Snooks, Greame Donald (ed.). *Was the Industrial Revolution Necessary?* London: Routledge, 1994.

Starke, Linda (ed.). *State of the World 2009*. New York: W.W. Norton and Co., 2009.

Steele, David Ramsay. *From Marx to Mises*. La Salle, Ill.: Open Court, 1992.

Stern, Nicholas. *The Economics of Climate Change: The Stern Review*. Cambridge, UK; New York: Cambridge University Press, 2007.

Stern, Nicholas. *Why Are We Waiting?: The Logic, Urgency, and Promise of Tackling Climate Change*. Cambridge, MA: The MIT Press, 2015.

Strotz, Robert H. "Two Propositions Related to Public Goods," *Review of Economics and Statistics* 40 (1958): 329–331.

Stroup, Richard L. and Goodman, Sandra L. "Property Rights, Environmental Resources, and the Future," *Harvard Journal of Law and Public Policy* 15,2 (Spring 1992): 427–454.

Tverdek, Edward. "The Efficiency Imperative: Five Questions," *Science & Society* 68,4 (Winter 2004–2005): 447–474.

Wenz, Peter S. *Environmental Justice*. Albany: State University of New York Press, 1988.

Willey, Zach. "Behind Schedule and Over Budget: The Case of Markets, Water, and Environment," *Harvard Journal of Law and Public Policy* 15,2 (Spring 1992): 391–425.

Willis, K. G. and Corkindale, J. T. (eds.). *Environmental Valuation: New Perspectives*. London: CAB International, 1985.

Wittgenstein, Ludwig and Anscombe, G. E. M. *Philosophical Investigations: The German Text, with a Revised English Translation*. 3rd edition. Malden, MA: Blackwell Publishers, 2001.

Wrigley, E. A. 1994. "The Classical Economists, the Stationary State, and the Industrial Revolution," in Snooks (ed.), *Was the Industrial Revolution Necessary?* London; New York: Routledge, 1994, 27–42.

Yandel, Bruce. "Escaping Environmental Feudalism," *Harvard Journal of Law and Public Policy* 15,2 (Spring 1992): 517–539.

Index

About the Author

Edward Tverdek is a Franciscan brother currently in studies for priesthood at Catholic Theological Union in Chicago. He teaches philosophy at Quincy University and is the author of numerous articles and reviews addressing areas as diverse as environmental policy, the nature of privacy, Marxist theory, the philosophy of religion, and feminist theology in journals such as *Science & Society, Public Affairs Quarterly,* and *The Australian eJournal of Theology.*

www.ingramcontent.com/pod-product-compliance
Lightning Source LLC
LaVergne TN
LVHW101445200926
848483LV00001B/57